GEOENGINEERING, PERSUASION, AND THE CLIMATE CRISIS

GEOENGINEERING, PERSUASION, AND THE CLIMATE CRISIS

A GEOLOGIC RHETORIC

EHREN HELMUT PFLUGFELDER

THE UNIVERSITY OF ALABAMA PRESS | TUSCALOOSA

The University of Alabama Press
Tuscaloosa, Alabama 35487-0380
uapress.ua.edu

Typeface: Minion Pro

Cover image: "Geographie der Pflanzen in den Tropen-Ländern," 1807, by Alexander von Humboldt; collection of the Leibniz-Institut für Länderkunde, Leipzig.
Cover design: Michele Myatt Quinn

Cataloging-in-Publication data is available from the Library of Congress.
ISBN: 978-0-8173-2142-0
E-ISBN: 978-0-8173-9424-0

To my parents, Helmut and Elissa Pflugfelder

CONTENTS

FIGURES

ACKNOWLEDGMENTS

During the course of researching and writing a book, lots of things happen. Kristin Griffin, my partner in all things, has been there from the first ideas to the final copyedits. I love you more each day, as we continue to figure out life together. That includes a life with Penny—nineteen pounds of funny, happy, and scrappy malshipoo. We are family. That family also includes my parents, Helmut and Elissa, my sister, Heidi, my brother-in-law, Brian, my parents-in-law, Lynne and Tom, brother-in-law, Matt, sister-in-law, Kim, and nephews, Luke and Chase. Thanks for being supportive and loving in all that you do.

My rhetoric colleagues at Oregon State University have always been there to help me sort the good ideas from the bad. Thanks especially to Tim Jensen, Lisa Ede, Ana Milena Ribero, Sarah Perrault, Vicki Tolar Burton, Chris Anderson, Anita Helle, Kristy Kelly, Dennis Bennett, Joshua Reeves, and Chelsea Graham. I owe you all. Also, I don't think I would have started this project had I not been invited to talk about geoengineering with students in Jake Hamblin's Anthropocene-themed honors course—I know you said the topic was my idea, but I honestly think you suggested it. We may never know the full story. Many more people at Oregon State deserve thanks; some of you have had great research suggestions, others offered great conversations, and still others extended great life advice over cocktails or chance encounters in the woods. Thanks to Lily Sheehan, Chris Nichols, Ray Malewitz, Peter Betjemann, Felicia Phillips, Rich Collins, Clare Braun, Tekla Bude, Neil Davison, Emily Elbom, Molly McFerran, Rebecca Olson, Liz Delf, and Megan Ward.

I also want to recognize all those with whom I have collaborated over the past few years. Thinking with you and writing alongside you all has given me more insight that you may know. Thanks to Dan Richards, Tim Amidon, Jen Clary-Lemon, Alex Mahmou-Werndli, Shannon Kelly, Ruth Sylvester, Marisa Yerace, Matt Fuller, Alex Nielsen, Sonia Stephens, Kristen Moore, Michael Salvo, Joshua Prenosil, Christopher Marshall, Evan Baden, Peter Konstantinidis, Melanie Link-Perez, Dave Rieder, Lydia Wilkes, Shawn Stowe, Bill Hart-Davidson, Jeremy Cushman, Randall Monty, Amber Buck, and Stephanie Vie. Beyond researching and presenting, some of the most rewarding collaborations I've been a part of in the past decade have involved editing and preparing issues of *Present Tense*. That crew of amazing and talented individuals includes Megan Schoen, Cristyn Elder, Don Unger, Ryan Skinnell, Matt Cox, Shreelina Ghosh, Liz Angeli, Jess Clements, Caitlan Spronk, Allen Brizee, Alex Hidalgo, and John Pell.

Finally, no acknowledgments would be right without thanking my editor at the University of Alabama Press, Dan Waterman. From the first conversation we had about this project, to handling reviews and assuaging my concerns, you've been fantastic to work with. Thanks, too, to Joanna Jacobs and Irina du Quenoy, wonderful editors and wordsmiths. Finally, thanks to the anonymous reviewers. Though I can only guess who you are, I appreciate your candor and honesty.

INTRODUCTION

Geoengineering and Persuasion

We should start this enquiry into geoengineering by first heading back in time—all the way back to Earth's most infamous extinction event. Around sixty-five million years ago, three-quarters of all living plant and animal species died in an episode immortalized both in the geologic record and in the pages of hundreds of textbooks. Most often depicted with a Tyrannosaurus rex squinting skyward at a massive asteroid about to plunge the earth into a hellscape from which few species would survive, the Cretaceous-Tertiary extinction event, also known as the Cretaceous-Paleogene extinction event, marked the end of the dinosaurs and the emergence of mammals as the dominant class of animals. When it impacted the northern edge of the Yucatán Peninsula, the asteroid largely responsible for this extinction vaporized everything within a thousand miles, set off tsunamis of unimaginable size, created winds of six hundred miles per hour, caused worldwide earthquake events, and expelled enough debris to block out the sun for the better part of a decade. The earth was plunged into a period of extreme darkness, increased atmospheric CO_2, and ocean acidification, thereby killing off many of the species that survived the initial impact. It also appears likely that the asteroid's collision intensified volcanic activity around the globe, with massive eruptions occurring in the Indian subcontinent, where lava flows now known as the Deccan Traps covered half a million square miles.

Part of why the end-Cretaceous mass extinction event is so widely known is because of its fantastic drama: a world full of giant dinosaurs that no longer exist, an extreme occurrence almost beyond our capacity for imagination, and a rapid change in speciation that left mammals to inherit the earth. Such disasters are mythologized in part because they are so extreme and unusual, but incidents like the Cretaceous-Tertiary extinction event are rare—far less common than the ebbs and flows of other causes of extinction. Fluctuations in CO_2 bring glaciers from the poles to Earth's currently steamy equator and back again; ocean acidifications cause the death of coral reefs, leave shelled invertebrates without protection, and fill the oceans with anoxic dead zones; increased temperatures reduce rainfall and starve terrestrial plants and animals. The dramatic Cretaceous-Tertiary extinction event has also been mythologized because there is something fundamentally comforting to modern-day humans in thinking that it would take the impact of a fifty-mile-wide asteroid traveling at sixty-five thousand miles per hour to change the conditions that account for our current widespread planetary habituation. For some, only a dramatic deus ex machina of such magnitude could alter the climactic conditions we currently enjoy.

The fact is, we have been enjoying a comparatively stable period in Earth's history. Relatively low atmospheric CO_2, few major volcanic events, and a moderate climate have remained steady throughout the period known as the Holocene epoch—to the extent that human experience has been one of relative climactic constancy. Beginning approximately eleven thousand years ago, after the last major glacial period, the Holocene epoch also afforded us an exceptional way of thinking, where our beliefs became grounded upon the knowledge that the world is stable and unchanging and will be for the foreseeable future, barring some unexpected extraterrestrial collision. We commit a what can be termed a *Holocene fallacy* when we believe in the power of this stability and simultaneously misunderstand or ignore how geologic forces can change our living conditions. A Holocene fallacy offers license to discount how both the deep history of the earth and ongoing geologic circumstances impact our lives, while also reinforcing the split between the sciences and the humanities. It offers us a way to neglect how millions of years of gradual climactic changes power, quite literally, almost all modern features of Western civilization. Perhaps most important, however, our current Holocene stability allows us to disregard our own ability to change the earth's climactic features through the abuse of carbon fuel accumulations from past geologic events. In short, by succumbing to a Holocene fallacy, we misunderstand the power of the geologic to continue changing the earth.

As a result of this general ignorance and the continual, almost unabated, extraction and ignition of past geologic carbon stores, we are now creating climactic conditions that threaten the very Holocene stability we have been enjoying. Current atmospheric CO_2 levels are greater than they have been since the Pliocene epoch, about three million years ago. In the Pliocene, sea levels were ninety feet higher, grasslands dominated, and much of the world was an arid desert. Some will respond to abrupt, negative, and alarming climactic conditions by continuing to deny that our stability can be permanently ruptured, but for others, including many leading atmospheric scientists, entrepreneurs, and increasing numbers of politicians, the answer to climate change could be through intentional climactic manipulations whose scale of implementation rivals these geologic forces. For some, the answer to this instability lies in geoengineering.

Geoengineering is a word like "television," "automobile," or "sociopath," a hybrid fused from Greek and Latin. The name connects the Greek term for the "earth" with "engineer," a Latin word meaning to arrange, contrive, or manipulate.[1] This linguistic mashup is especially fitting, as the term denotes complex scientific and engineering strategies that intend to change some aspect of the earth on a large scale, with an end goal of altering the earth's entire climate system. Geoengineering strategies can be traced back to weather modification efforts made by scientists and governments throughout history.[2] The first explicit use of the term might come from physicist Cesare Marchetti's 1977 article "On Geoengineering and the CO_2 Problem," in which he suggested that industrial CO_2 could be piped under the world's oceans.[3] The later and more influential Royal Society report *Geoengineering the Climate: Science, Governance and Uncertainty* defines geoengineering (which also goes by the names climate engineering or climate management) as the "deliberate, large-scale intervention in the Earth's climate system, in order to moderate global warming."[4] These intentional climate-altering projects hypothesize how to reduce or reverse the rate of temperature change, remove greenhouse gases (GHGs) from the atmosphere, or otherwise alter earth's relationship with greenhouse heating from the sun's rays.

Geoengineering strategies can be placed into one of two categories, though there are dozens of different versions. The first method is one that many Americans have likely heard of, if they have ever encountered "clean coal" technology. These are carbon dioxide removal (CDR) strategies, and they sometimes go by the name "carbon engineering," or as Clive Hamilton has often described it, "sucking carbon." In brief, these strategies involve the reduction of carbon, either at the source of pollution, like in clean coal, or through various filtering processes that absorb carbon from the atmosphere, as in direct air capture (DAC). Some CDR strategies, like enhanced weathering, desert flooding, or massive new tree farms, are relatively benign, but in others, where carbon is pulled from the air through industrial means, that carbon needs to go somewhere once extracted. This is often where things get thorny because most chemical manipulations of airborne carbon involve storing CO_2 as a mineral, injecting it into oil fields or coal seams or forcing the ocean to absorb more CO_2. That latter idea is looking less like an attainable solution, as the world's oceans become increasingly less able to store excess carbon dioxide. All said, though, CDR carries less risk than the other major form of geoengineering but also has less immediate impact on atmospheric CO_2 and is therefore less effective in terms of reducing global temperatures.

The second major kind of geoengineering is called solar radiation management (SRM), also known as "regulating sunlight."[5] These plans do not try to pull carbon from the atmosphere but instead work to reduce the effect of the sun's rays upon the earth. They either change how the earth absorbs the solar rays or try to alter surface albedo—the reflection coefficient of the earth, where a coefficient of "0" indicates no reflectivity and "1" indicates perfect reflection. A more reflective Earth, for example an Earth covered by additional ice or clouds, invites cooler temperatures because the atmosphere retains less heat. Geoengineering schemes that focus on blocking solar radiation run the gamut from the mundane (increasing the use of light-colored roofing materials) to the complex (artificially brightening marine clouds or spraying mountains with a mix of water, sand, and lime) to the audacious (partially blocking the sun by way of millions of tiny atmospheric mirrors). The hope is that with enough sunlight blocked and reflected, the earth will cool by several degrees and allow enough time for other carbon-management activities to take effect. When they do, and the level of CO_2 in the atmosphere reaches acceptable levels (somewhere between the 280 ppm [parts per million] from before the Industrial Revolution to the 419 ppm we have as I write these words),[6] we can then slowly dismantle whatever sunlight regulation practices we have in place. SRM is often riskier, but hypothetically cheaper, and could have more of an immediate impact on temperature.

In order to wrap our heads around geoengineering a bit more, we should consider one specific approach. One of the more alarming-sounding geoengineering plans is also one that many scientists and policy advocates are interested in pursuing: an SRM strategy that looks to mimic the effects of a massive volcanic eruption, which requires a little backstory. When Mount Pinatubo erupted in the Philippines back in 1991, the explosion killed hundreds of people, spewed more than ten billion tons of magma, and, most importantly for climate scientists, released twenty-two million tons of sulfur dioxide (SO_2) into the atmosphere. What at the time was a humanitarian crisis and nightmare for regional agriculture became useful data for scientists studying atmospheric change. The eruption was so large that it released an aerosol cloud of SO_2 that spread around the globe in the following months. The next year, the eruption cooled the Northern Hemisphere "up to 0.5 to 0.6°C, equivalent to a hemispheric-wide reduction in net radiation of 4 watts per square meter and a cooling of perhaps as large as -0.4°C over large parts of the Earth."[7] Basically, the eruption produced a cloud of atmospheric particles dense enough to block out a significant portion of the sun's rays, initiating a global cooling event. Geoengineering supporters see this as evidence that a similar event, popularly called the "Pinatubo option," could be intentionally instigated and maintained as a way to offset global warming; aerosol particle delivery would be tricky but could be achieved by high-altitude aircraft or perhaps a giant hose. These same proponents of a stratospheric particle injection (SPI) strategy claim that no major advances in technology would be required for the delivery of sulfur aerosols and that the plan would impact global temperature in less than a year. The proposal is likewise relatively cheap, supposedly costing about $150 million to launch and $50 million per year—a claim that received widespread popular exposure when it was repeated in Steven Levitt and Stephen Dubner's in *SuperFreakonomics*.[8] Unsurprisingly, the potential downsides to such a plan are extreme and include effects such as additional ocean acidification and the disruption of wind patterns and rainfall events, including the possible cessation of African and Central Asian monsoons and a consequent disruption in the food supply for billions of people. Unfortunately, as the following chapters show, SPI is also set to dominate future deliberations about our planet's climate.

Geoengineering schemes like these sound extraordinary but increasingly are no longer thought of as simply fringe science. A number of reputable and progressive climate scientists are taking geoengineering seriously and have begun the process of researching specific geoengineering technologies. Paul Crutzen, the Nobel Prize–winning atmospheric scientist who discovered the chemical reactions that led to the discovery of the ozone hole, is cautiously optimistic about a Pinatubo option. In 2006 he coauthored the paper "Albedo Enhancement by Stratospheric Sulfur Injections: A Contribution to Resolve a Policy Dilemma?", breaking the modern-day taboo in scientific circles on the topic of geoengineering. Likewise, astrophysicist and America's most prolific patent holder Lowell Wood is in favor of sulfate aerosol SRM, celebrated engineer Stephen Salter has suggested artificial cloud albedo enhancement, Harvard physicist Russell Seitz believes microbubbles in the ocean could enhance surface albedo, and even dedicated environmentalist James Lovelock has advocated for giant ocean pumps thought to accelerate CO_2 transfer.[9] But these concepts are not just hypothetical thought experiments dreamt up by a few climate scientists—wealthy philanthropists and businessmen such as Bill Gates, Richard Branson, and Niklas Zennström (the cofounder of Skype) have been funding geoengineering research. Gates has helped fund SRM plans and Branson offered a $25 million prize for geoengineering ideas. Nathan Myhrvold, former chief technical officer at Microsoft, has used his company Intellectual Ventures to explore several geoengineering concepts, including a solar shield and a plan similar to Lovelock's.[10]

Never one to be outdone, Elon Musk announced in 2021 that he would donate $100 million dollars toward carbon capture and storage. In the United Kingdom, researchers from Bristol University, Cambridge University, and Oxford University formed the SPICE project. Not a pop band, this SPICE is short for Stratospheric Particle Injection for Climate Engineering, and scientists at the three schools have been researching SRM techniques and the possibility of testing atmospheric aerosol injection (they had a planned test canceled in 2012, purportedly due to a conflict of interest).[11] We are also seeing funds for geoengineering research earmarked in governmental climate plans. In 2020 presidential candidate Andrew Yang claimed he would dedicate $800 million to geoengineering research and the Democrat-led Select Committee on the Climate Crisis recommended funding additional atmospheric climate intervention research as part of a larger risk management strategy for climate change.[12] To put it briefly, geoengineering is now being taken seriously by a wide range of scholars, scientists, philanthropists, and politicians, and it is increasingly likely to play a large role in discussions about global warming in the coming years.

As a debate over what we could or should do, geoengineering is an example of deliberative rhetoric. Deliberative rhetoric concerns reasoning and discussion over future actions and the evaluation of those actions as good/bad or advantageous/disadvantageous. For Aristotle, deliberative rhetoric was germane to the legislative assembly, where "the deliberative orator is concerned with the future: it is about things to be done hereafter that he advises, for or against."[13] While Aristotle identified five possible deliberative matters ("ways and means, war and peace, national defense, imports and exports, and legislation"), deliberation over geoengineering could encompass all five of these concerns and more.[14] A choice that will affect the future of the entire world, geoengineering represents one of the most significant deliberations of the twenty-first century, because the question of whether we should implement geoengineering is bound up with the world's reaction to climate change, when those responses could be put into action, and who would have the right to realize them. Current rhetorical strategies in support of geoengineering are dominated by claims about appropriate timing and amplified through appeals to definitions of nature. It is through these strategies that the earth is being rhetorically weaponized to justify different perspectives on what kind of geoengineering we should research and when we should deploy the results of that research. Often geoengineering is positioned as a technological response to what Rob Nixon has called the "slow violence" of climate change, "a violence of delayed destruction that is dispersed across time and space, an attritional violence that is typically not viewed as violence at all."[15]

Yet geoengineering could easily become an additional form of this violence, as those with the ability to geoengineer could use it to solidify climactic conditions that have become favorable to Western powers. Complicating this deliberation is the rarity of public discussion about geoengineering—indeed, at present only a small percentage of people are aware of existing geoengineering proposals. Most serious conversations have happened within policy groups, behind the closed doors of climate-oriented start-ups, between subject matter experts at scientific conferences, or in the disciplinary jargon of research journals. Further, much of this dialogue has occurred solely in the West, and "so far, geo-engineering has been an almost exclusive debate in rich countries of the North."[16] Throughout this book, my concern is with bringing to light the rhetorical strategies of these discussions, experiments, and investments and criticizing their value, integrity, and impact.

On the whole, the current state of deliberation for and about geoengineering is deeply worrying, for several reasons. Too often, this discourse is pervaded by appeals to geologic conditions and claims to know what is "natural" or what has historically been part of always-ongoing earth processes. These appeals shift and morph depending upon the goals of the geoengineers, taking on multiple and varying definitions of what counts as "geologic." Some arguments suggest that geoengineering strategies can be as "natural" as a volcanic explosion and should therefore be employed to help the earth recover from human action. At other times, geoengineering proponents align their arguments with the wholly "unnatural" Anthropocene, emphasizing that since humanity has always changed nature, it should continue to do so but now with more knowledge and foresight. Still other geoengineering advocates seek out connections to free-market capitalism, in the belief that competition and investment will lead to innovative new climate solutions sure to improve and stabilize the planet's many ecosystems. These entrepreneurs sometimes radically simplify years of complex scientific research to make unjustifiable claims and generate a short-term profit for their investors. The risks of their real-world "experiments" are not yet well understood and tied to Western forms of paternalistic environmental stewardship, a strategy that ultimately reinforces a global status quo. Geoengineers are frequently guided by utilitarian scientific worldviews, and while they may prioritize humanitarian goals, many lack the ability to see the harm their propositions could cause.

Unfortunately, the rhetorical strategies coming from geoengineering advocates have been largely deceptive, hegemonic, deterministic, and exploitative. Philosopher Frederic Neyrat considers these discourses to be part of what he calls "geocontructivism," a nefarious framework that "considers the Earth as the consenting prey of an integral conquest."[17] He argues that geoconstructivism

> asserts itself at the intersection of several discourses: the discourses of engineers and architects who would like to transform the Earth into a pilotable machine; biologists who would rather spend their time resurrecting already extinct species rather than protecting those that are still alive; political strategists offering solutions for global governance; the advent of new markets by businessmen who view climate change as a new industry for economic speculation; geographers enthralled by the power of humanity within the age of the Anthropocene; sociologists and anthropologists proclaiming that there is no common world and so it is up to us to build one; essayists promoting nuclear energy for all; prophets declaring the death of nature or the birth of the transhuman; philosophers inviting us to accelerate our technological control over existing society; paradoxical ecologists simultaneously lauding the merits of fracking and dreaming of the disappearance of any form of ecology containing a political dimension.[18]

We need to understand the structure, content, purpose, and effect of these geoconstructivist discourses and sound an alarm about their deliberative directions. Rhetoric and communications scholars have a responsibility to understand how these arguments are made, what results from these persuasive strategies, and what harm they are causing and could cause. We need to make clear how geoengineering proponents intentionally and unintentionally marshal other geologic actors into their arguments and need to adapt our existing theories of rhetoric to study how these myriad concepts exploit and appeal to what I refer to as the "geologic." The rhetorical approach named in the title is a methodology attuned to the study of large-scale, Earth-altering persuasive strategies, and in the chapters that follow, I identify deliberative arguments for geoengineering, locate their origins, and critique the sometimes-pernicious ways they have been deployed. Many of these arguments are unsettling, and rhetoricians concerned with science and the environment need to scrutinize our current moment with greater intensity, as actions taken now could very well determine our climate future.

We should be especially cautious of the increasing pressure being put on global research agendas to understand geoengineering and where that research could lead. Geoengineering has the potential to become the default method for dealing with climate change, despite the staggering complexity and potential danger of its application. Political theorist Catriona McKinnon has argued that we are approaching what she calls a "cognitive lock-in to a more limited range of [geoengineering] options than is justified" and has voiced concern about the increased rate of research on SRM technologies in particular.[19] McKinnon maintains that ongoing research programs can render alternative climate change responses unlikely, because dominant research formations "can persist . . . even when there are many better ways to pursue the ends, or arrange the affairs, in question."[20] While McKinnon describes this framing as a "cognitive" lock-in, it is likewise accompanied by, and enabled through, the discursive and material persuasive forces that generate research momentum, encourage investment, and embolden real-world experiments in geoengineering as a solution to our ongoing climate crisis. Global inaction on climate change, combined with an inattention to the startling emergence of geoengineering, has meant that these persuasive formations are threatening to lock in discussions about geoengineering ideas even before most people have heard of them. It is therefore imperative that we study this deliberative formation more closely than we have. Subsequently, I establish the landscape of this deliberative discourse by introducing some of its basic rhetorical features before broadening out to consider a more geologically attuned approach to rhetoric.

TIPPING POINT EXIGENCE

The rationale for geoengineering is often founded upon the scientific consensus that we have already reached a more familiar kind of "tipping point" than McKinnon's. This tipping point refers to the release of GHGs, wherein we are now locked into impactful global warming scenarios. Claims that we are past the moment where carbon cessation could avoid destructive global temperature increases, and the attendant amplified climate volatility, also emphasize that we have reached a state of exception, one where global warming looms larger than any other environmental issue. These arguments stress that we are now in an exceptional scenario where (1) a slow drawdown of carbon emissions will not stave off the effects of global warming and (2) any further delays in acting upon global warming will simply compound the harm caused by the coming climate disaster. In response to this scenario, environmental groups often employ apocalyptic metaphors and make the case that aggressive and widespread negative carbon footprints, in all walks of life and in all industries, are necessary to avoid the very worst of global warming, now that the tipping point appears to have occurred.

What does this "tipping point" look like? We have reached, at the time I write this, a constant state of roughly 419 ppm of CO_2 in the atmosphere, a point that, based on current models for additional CO_2 output, will put the world between 2.4 and 6.2°C of temperature increase by the end of the twenty-first century (4.3 to 11.5°F).[21] This increase in temperature and the myriad climactic changes that will occur as a result are likely inevitable.[22] If we are able to reduce atmospheric CO_2 by phasing out all coal emissions by 2030, the most significant source of anthropocentric atmospheric CO_2, we could theoretically keep levels to where they are at present and slowly draw them back to 350 ppm. Of course, "even with phase-out of coal emissions and assuming IPCC [Intergovernmental Panel on Climate Change] oil and gas reserves, CO_2 would remain above 350 ppm for more than two centuries."[23] The CO_2 in the atmosphere, and therefore global temperature averages, are unlikely to drop for a thousand years following the cessation of emissions, which is not something that the word's countries and economies are ready to do, by any estimate. In their calculations, Damon Matthews and Ken Caldeira found that anthropogenic CO_2 emissions would have to be eliminated now (as in, 2007, the time of their article's publication) in order to stabilize current earth temperatures, a scenario that still remains so unlikely that some would call it nearly impossible.[24] The scientific work that has gone into defining and clarifying this tipping point has also created the exigence for geoengineering, especially as little progress has been made on meeting global GHG reduction goals.

Our increasing certainty about the irrevocable effects of GHG emissions feeds a growing anxiety that the world's human and nonhuman populations are in serious danger—a fear compounded by the lack of intergovernmental commitment to a management strategy that can fully confront that risk. Scientists of all stripes have been frustrated by this seeming incongruity and are increasingly willing to identify geoengineering as a "failsafe" from the worst of climate change-induced disasters. Jonas Anshelm and Anders Hansson have described the exigence for geoengineering as "the end of the world as we know it is approaching, and scientists are shocked by new scientific findings and observations. The alternatives, as they are presented, are either to inactively wait for the catastrophe or to explore the final option: geoengineering."[25] Increasingly, geoengineering solutions are represented as an unorthodox and alternative environmental strategy, that, while disliked and feared, has been proposed as a necessary evil. The atmospheric researcher Ken Caldeira has also argued along these same lines. "What happens," he asks, "if in 2040 or 2060 temperature increases are so high that crops are failing throughout tropical regions and billions of people are threatened with famine? We'd better try to understand if there is something we could do, because there's no other way to realistically stop the Earth from warming during the course of this century."[26] Like many topics within environmental discourse, geoengineering is being framed within an increasingly narrow window for action—a constraint caused by years of political in-fighting, despite overwhelming scientific agreement that climate change is already occurring.

In the midst of this expert consensus about climate change, many climate scientists are rightfully exasperated that governments have not as of yet acquiesced to meaningful carbon cessation agreements, strong controls on GHG emissions, or even recognized the scope and severity of the crisis. Certainly, evidence for anthropogenic global warming has been broadly known at least since James Hansen's statement on the greenhouse effect to the US Senate Committee on Energy and Natural Resources in 1988. But this awareness does not appear to have extended to legislators. In 2005 the US Senate approved a largely symbolic "sense of the senate" stating climate change is real and that humans contribute to it. Yet in 2015, they also voted down an additional amendment stating that climate change has been caused by human action.[27] In the face of overwhelming proof, longtime climate change denier Senator James Inhofe has maintained that the "climate is changing and climate has always changed and always will. . . . The hoax is that there are some people who are so arrogant to think they are so powerful they can change climate. Man can't change climate."[28] Further frustrations stem from the perceived failure of international agreements, including the Kyoto Protocol in 1997, the relatively ineffective 2009 Copenhagen Accord, the too-slow progress made by the United Nations Framework Convention for Climate Change, the lack of any binding mechanisms for the 2015/2016 Paris Agreement, and the milquetoast results of the annual meetings of the parties of the convention. This exasperation is best represented by James Hansen's reaction to the Paris Agreement and others like it: "It's a fraud really, a fake. . . . It's just bullshit."[29] Political denial and foot-dragging have left some climate scientists wondering whether intergovernmental agreements will ever identify or enforce any meaningful CO_2 reduction measures. The result of this frustration may turn some toward climate modification technologies that will halt or reduce global warming, if only because these strategies represent direct action. Significant numbers of scientists and science journalists seem to agree that politics have failed and that politicians have not acted upon the evidence for global warming that has been presented for the last thirty years. When continual economic growth is advocated over GHG reductions, this logic goes, geoengineering research becomes ever-more necessary.

A second, related frustration for the scientific community lies in the clash between scientific consensus and carbon-dependent industries. Powerful interests historically opposed to carbon cessation (and who have historically denied anthropogenic global warming) can be grouped together as "the carbon lobby." This term encompasses individual corporations that produce fossil fuels, including the twenty companies who release a third of all GHGs, but also industry lobbying groups such as the American Coalition for Clean Coal Electricity, the International Emissions Trading Association, the European Chemical Industry Council, and the International Air Transport Association, to name a few.[30] These groups have stymied efforts toward a carbon-neutral economy and have actively lobbied against carbon-neutral (or -negative) technologies. As Alex Steffen explains, "for the last thirty years, the Carbon Lobby's strategy on climate change has been to delay," for the simple reason that delaying carbon reduction means continued sources of profit.[31] Such delaying tactics have been taken up by others, manifesting in several obstructionist strategies. Lamb et al. found four main approaches to these "discourses of delay": (1) redirect responsibility, (2) push nontransformative solutions, (3) emphasize the downsides, and (4) surrender to climate change.[32] It takes very little imagination to evaluate the current environmental and political impasse and conclude that the only way out is through a radical, technological (and supposedly nonideological) solution that avoids these longstanding political and industrial bottlenecks.

Anthropogenic global warming is happening, atmospheric CO_2 is the driving factor, and exponential economic growth is currently at odds with carbon cessation, yet because of legislative inaction, we face a situation where any slow draw-down of carbon emissions is not going to halt the effects of climate change. Climate scientists have been vocal about the dangerous consequences of rising CO_2, with great consensus, but lobbyists and governments seem to favor slow, incremental change (if any) to an increasingly urgent problem. For geoengineering supporters, these details result in the assumption that any strategy that could mitigate the effects of climate change should be heavily researched. If we can understand and perhaps employ geoengineering while continuing to pursue carbon-neutral technologies, we might avoid the worst climate change scenarios. It is in this milieu that geoengineering is framed as a necessary evil that can "buy us time" until many of the world's economies can transition to no- or low-CO_2 emission industries. Partially because of these stress factors, a curious sort of coalition has developed over geoengineering.

A STATE OF EXCEPTION

One of the best-understood effects of global warming is the calamitous warming and acidification of the world's oceans, a disorder putting many different animal species at risk. In this regard, one of the most widely discussed and reported-upon impact sites has been the Great Barrier Reef. Australia's massive coral reef in particular has suffered from several waves of bleaching events, wherein water that is too warm forces coral to eject their resident algae, causing the coral to turn white and jeopardizing their life.[33] Such is the attention that the Great Barrier Reef has received that when the Reef Restoration and Adaptation Program (RRAP), a coalition of Australian universities, parks, research agencies, and charities, announced their geoengineering efforts, the plan was widely welcomed. RRAP included $10.4 (AUS) million for removing coral-eating crown-of-thorn starfish, $36.6 million to divert polluted water from entering the reef, and additional funds toward experimental work described as "'brightening' marine clouds to shade reefs."[34] This fairly innocuous language went largely unnoticed by the press until early 2020, when researchers tested cloud-brightening equipment 100 km off the coast of Australia. The team experimented via a process wherein microscopic sea water droplets were sprayed into the air, allowing salt particulates to seed marine clouds. The assumption is that additional clouds could form a protective layer above the Great Barrier Reef, reflect harmful sunlight, cool the water, and (hypothetically) preserve existing coral. This research consortium plans to continue testing this system, hoping to increase the size of these experiments, "which the researchers say should be able to brighten clouds across a 20-by-20-kilometre area."[35]

Perhaps most noteworthy about this test is not that it was reported to have been successful, but that it attracted very modest attention outside of a few articles by geoengineering watchdog groups.[36] Some of this lack of attention was certainly due to the emergence of the COVID-19 pandemic, but cloud-making is also seen as relatively benign when compared to other geoengineering methods. While SRM experiments that mimic volcanic eruptions have raised concern (hence the canceled 2012 SPICE experiment and the delayed Stratospheric Controlled Perturbation Experiment (SCoPEx) studies, the creation of additional marine clouds has been generally perceived as nonthreatening. Additionally, RRAP's research is part of a strategy to save what is considered a highly valued biological treasure. A World Heritage site and marine park, the Great Barrier Reef is frequently used as a metonym for ocean health in general, which, when combined with a less-than-dangerous-sounding geoengineering plan, allowed this instance of cloud-based geoengineering a "pass." Finally, and perhaps most disturbingly, RRAP's test failed to raise many alarms because it did not address the volume of CO_2 that has already been released and therefore does not combat the key compound in global warming. The forms of geoengineering that do not address CO_2 are especially concerning, because they are fueled not only by a growing a sense of urgency but also by a surprising and persuasive consensus.

Geoengineering deliberations appear to be flourishing in what some have called a "state of exception," where some tacit agreements have occurred between frustrated climate scientists, conservative politicians, conservative think tanks, and the fossil-fuel energy industry. Specifically, these groups all appear to concur that geoengineering should be better understood, researched accordingly, and possibly deployed in the near future. The reason for this strange alliance is that some influential climate scientists who no longer believe that human efforts toward carbon cessation will happen in time are beginning to speak the same language as the carbon lobby and those who wish to profit from climate management technologies. Geoengineering projects, with their large scope, established need, and potential impact, also represent a financial opportunity for some. There is room to patent specific techniques, develop novel technologies, bolster existing industries, and ultimately form a new industry around the central tenet that geoengineering will eventually become a profitable enterprise. The emergence of companies like Intellectual Ventures, Carbon Engineering, Climos, and Planktos Ecosystems, among dozens of others, attest to this recognized prospect. Tina Sikka has argued that recent high-profile investment from tech luminaries like Bill Gates "stifles serious debate on the deployment of such technologies" by creating the appearance of legitimacy of a "technological and market-driven fix to climate change."[37] Other critics like Bronislaw Szerszynski caution that "we should not be naïve about the significant economic interests which would thereby become involved" in geoengineering, nor should we be surprised that SRM projects become part of the "neoliberal project to extend the use of market mechanisms into more and more areas of economic and social life."[38] These market pressures and the unlikely alliances built around them have the potential to become more powerful as climate change intensifies.

Scholars in the social sciences have been concerned about this conciliation for some time. Eric Swyngedouw has explained that discourse about climate change and its growing state of exception offers additional evidence for the breakdown of traditional political perspectives. He argues that "in the domain of environmental policies and politics, [there is] a widespread consensus that Nature and the Environment need to be taken seriously, and that appropriate managerial technological apparatuses can and should be negotiated to avoid imminent environmental catastrophe."[39] Swyngedouw's assessment of this "post-political" climate change discourse echoes the work of Alain Badiou and others but brings the discussion back to the neoliberal rationale for state-controlled environmental measures.[40] He argues that "in the name of saving the planet all manner of expertise, knowledge, technologies, and arguments are marshaled to perform their prophylactic qualities," where our current "condition is increasingly characterized by depoliticizing modes of governance and increasingly repressive forms of bio-political control."[41] This depoliticized "state of exception" applies to an environmental politics that embraces sustainability and utility, in which any disagreement from the call to consensus becomes tantamount to heresy. For advocates of geoengineering, this environmental exceptionalism can be mounted by experts and political advocates "to present it as the only option with respect to mitigating catastrophic global warming."[42] Such an exceptional status could emerge because we have reached a juncture in the global release of GHGs where the effects cannot easily be undone and extraordinary measures will be needed to maintain existing states of nature and current standards of livability (such as they are). After all, who would argue that the survival of one of the world's most beautiful and lauded ecosystems—like the Great Barrier Reef—should not be saved through the generation of a few additional clouds? Who would argue against implementing geoengineering if it could save the world's low-lying island nations, like the Maldives or the Marshall Islands? Likewise, who would argue against geoengineering, if research has conclusively shown its large-scale effectiveness in combatting the most dangerous effects of climate change?

Advocates for geoengineering often present it as a reasonable option, and possibly one of the only realistic options, in a world where the consequences of climate change are being made manifest, regardless of current abatement efforts. Unfortunately, as part of an argument for immediate action as opposed to gradual policy, industry, and lifestyle change, geoengineering discourse can perform a chilling effect on traditional political divisions. In this regard, Nigel Clark sees the emergence of an atmosphere of crisis where "there is never enough time for proper deliberation or collective consultation" about how to implement comprehensive changes.[43] In this discourse, deliberation shifts to what kind of additional dangers each geoengineering plan creates and what collateral damages are worth accepting. Any explicitly "politicized" perspective on geoengineering is thus suspect, because if a group completely opposes it, their motives are assumed to be in misalignment with the broader environmental values directed at "saving" much of the earth. Science writer Eli Kintisch suggests that under this exceptionalism, a strong critique or distrust of geoengineering strategies (because they are unnatural, too risky, or an anathema to the goals of genuine sustainability) is evidence, for conservatives, that liberals "don't really want to solve the problem—or even that they have more ulterior motives."[44] The default position on climate change could become, as the conservative American Enterprise Institute explains, that "geoengineering can stave off short- and medium-term harms while giving time for a long-term solution to take effect," thereby resulting in "a coherent policy proposal that may enjoy broader public support."[45] That is, the issue could become not *whether* to research and deploy geoengineering but *which* strategy to adopt and *when* to implement it. Tina Sikka summarizes this rhetoric of exceptionalism as "almost hegemonic" in the way that it presents "our current choices in such stark terms, i.e. between catastrophic disaster and life-saving technical intervention [where] using the trope of 'the tipping point', the 'right' decision, ipso facto, is made for us."[46] Complicating this situation further is the difficulty of achieving true international consensus on geoengineering, because politicians, activists, researchers, and investors define the conditions for action in strikingly dissimilar ways. Further, what might be good for one country (geoengineering could lower temperatures and keep the American Midwest from turning into a permanent dust bowl) could be terrible for another (worsening global warming could extend the Siberian growing season). What may represent "right timing" for geoengineering deployment for some could lock future generations into a foreign-controlled temperature-control technology. It will come as no surprise to rhetoricians, then, that "right timing" is also at the heart of any geoengineering discussion.

GEOENGINEERING KAIROS

Kairos is an easy enough concept to locate within the logics of geoengineering advocacy, as almost all rationales situate these strategies as methods for "buying time" until a more permanent solution to climate change can be found. Kairos is the companion to chronos, where kairos is often translated as "right timing" and a qualitative evaluation of time, setting, and audience used to determine when an action should be undertaken. Kairos can be defined most succinctly as "propitious timing," though as Debra Hawhee clarifies, it is not a kind of "right timing" that the author, or rhetor, can fully control.[47] That is, kairos is not dictated by the setting of a rhetorical act any more than it is determined by the rhetor alone. Hawhee points out that while scholars such as James Kinneavy and James Baumlin identified kairos as existing outside the body of the rhetor and "grounded primarily in rationality and reasoned principles" of accommodation and creation, kairos is much more immanent and produced.[48] What constitutes "right timing" to a specific situation is difficult to understand without having deep knowledge of, or being immersed within, the material conditions where timing is being evaluated. It can help if we think about kairos within something like a soccer game, where the "right timing" for a goal is generated by the players as much as it is by their training schedules, the rules of the game, the pitch, the weather, the crowd, and a whole host of other factors, each contributing to the moment when a play comes together. Therefore, we might say that kairos can never exist "on its own," apart from the perceptions or actions of an individual, any more than an opening in a particular moment of play of a game exists independently of the positions and skills of the players."[49] Kairos is immanent to the material and symbolic forces that must necessarily host the change for propitious timing to occur.

Certainly, any rationale for future geoengineering decision-making includes a sense of kairos as key to a geoengineering strategy. Different approaches require unique stages of planning and enroll entirely different groups of people and technologies into an evaluation of timing. For example, we can consider what a population of frustrated climate scientists might believe about geoengineering, accounting for the material and discursive conditions of the world in which they have produced knowledge. On the whole, they overwhelmingly agree that increased, anthropogenically caused atmospheric CO_2 is causing climate change. They might be largely in favor of implementing some of the more benign CDR technologies—those that pull carbon from the air—though they may also be dismayed by their relatively high implementation cost and their slow rate of effectiveness. Because CDR techniques reduce the amount of CO_2 in the atmosphere and/or oceans and appear to entail fewer negative environmental side-effects than other options, they are assumed to be currently deployable, with further research needed to improve upon existing techniques. For this group, the kairotic moment for CDR could be now, or perhaps thirty years ago, and the sooner functional CDR technologies come online, the better. On the other hand, SRM technologies—those that block the sun—might be seen as emergency measures that we should have at the ready, in case they are warranted, but not ones that we should immediately adopt. SRM strategies are seen as less desirable by climate scientists, because they do not change the level of CO_2 and come with a wider range of detrimental side effects. Our hypothetical group of climate scientists might advocate for additional research on SRM but would be much more cautious as to their implementation, because they could be needed in the future if their implementation becomes urgently kairotic.

For conservative think tanks and the carbon lobby, a different logic of kairos prevails. Because these groups are relatively uncertain (or are calculatingly obtuse) whether increased CO_2 contributes to global warming, the right timing for the deployment of SRM technologies is unlike that which is manifest for climate scientists. For them, a kairotic moment for deployment does not depend on averting a humanitarian or environmental catastrophe, but on an economic motive. The ongoing delay in curtailing CO_2 emissions is directly linked to maximizing advantages from the current exploitation of fossil fuels while waiting for a suitably propitious moment to invest in alternative fuel sources. This postponement is largely due to what has been called "the benefits of procrastination," a kairotic strategy employed by those who have generated economically oriented CO_2.[50] The logic behind such procrastination works like this: by allowing high CO_2 emissions to continue, the fossil-fuel industry is sustained, and GDP remains high, whereas cutbacks in energy use or support for alternative energies are generally assumed to hurt Western economies (though this could change in the future). If and when atmospheric CO_2 leads to economically disadvantageous global warming, then patentable or otherwise proprietary SRM strategies could be deployed to curtail temperature rise and the benefits of current geoengineering research would finally "pay off." While many might find this procrastination strategy to be a horrifying and wholly unethical formation of kairos, for one group it allows for maximum economic gain that is apparently worth the negative consequences.

Yet, regardless of the approach toward "right timing," both CDR and SRM forms of geoengineering enlist future generations in any evaluation of kairos. Or to put it another way, the "impact lag" of any geoengineering strategy is substantial. Because CDR technologies pull CO_2 from the atmosphere at a slow rate but can be stopped at any point, they place the burden of removing the bulk of atmospheric CO_2 onto future generations, without locking them into one specific geoengineering plan. Conversely, SRM technologies are predicted to have a greater immediate impact on global temperatures (SPI plans in particular). And while they also shift the burden of removing atmospheric CO_2 to future generations, they commit those generations to an individual geoengineering strategy, because turning off SRM technologies often means a return to rapid and disastrous global warming.[51] Either way, both climate scientist- and carbon lobbyist–led deliberations leave future generations to inherit the burden of their current calculations. Such is the impact of our current deliberative moment that "right timing" is always a calculation that enrolls future generations into a specific regime of consent.

A GEOLOGIC RHETORIC

As one might imagine, the suite of outlandish-sounding concepts called geoengineering have so far been a difficult one for many rhetoric scholars to examine (with Sikka's research excepted). Investigations into geoengineering are at a stage where increased knowledge about these projects is just now emergent; aside from Cold War–era experiments on weather manipulation and outside of science-fiction, much of the research conversation on geoengineering is less than two decades old. However, it is imperative that we interrogate these persuasive strategies, so as to better understand the power relationships being generated in our current deliberations. Part of this practice involves recognizing how these arguments are being made, something that encompasses more than a scrutiny of the language being used in scientific journals, opinion pieces, or think tank position reports. Because geoengineering plans will call into question the whole earth system and affect everything living and nonliving on the planet, any framework we use to understand the persuasive structures of geoengineering advocacy should take into account both symbolic elements as well as nonsymbolic, nonhuman forces. A rhetorical study of geoengineering should build from recent work in material rhetoric that can see complex ecologies of persuasion, and at the same time avoid making homogenous assumptions about people and their environmental conditions and histories. To be useful for studying geoengineering projects, and perhaps the Anthropocene in general, rhetorical analysis should deal explicitly with how geologic and atmospheric forces impact persuasion. While rhetorical theory has welcomed new material forces to the study of persuasion, it has so far inadequately dealt with the immense, primordial geologic and atmospheric forces that have shaped and continue to shape the very ground we live on and the ontological grounds for our known ways of being.

Materially oriented rhetoric has been advanced by studies in activity theory, actor-network theory, new materialisms, Indigenous ontologies, and object-oriented philosophy. New materialism in particular has opened up ways of refiguring the relationships between things themselves—not only regarding human-to-human interactions. Jane Bennett's work, notably her explication of vital materialism, has been frequently employed to understand the latent persuasive rhetorical agencies emergent in different assemblages and groupings of actors,[52] yet material rhetorics have also used other theoretical foundations for examining how objects can be persuasive.[53] With this influx of new enquiries, and additional critiques from Indigenous scholars, rhetoricians have explored the role of persuasion in complex discursive-material relationships, from the multiple physical interactions inherent in the process of writing to environmental and ecological conditions in studies of sustainability, to various other object relations in the world. This work has questioned rhetoric's epistemic assumptions and the implications of those assumptions for agency, intentionality, and power. New materialist rhetorical arguments have also successfully expanded rhetoric's focus beyond the human and beyond intentional, human-oriented symbolic persuasion. While this work has focused less—if at all—on geologic and atmospheric forces, it has thoughtfully and purposefully investigated the intersections between the material and the ecological.

In *Tracing Rhetoric and Material Life: Ecological Approaches*, Wells et al.

describe rhetoric's increasing attentiveness toward materiality and ecology.[54] They argue that while new materialist approaches have created room for theorizing about the many material entanglements of the world, it would be a mistake to simply reclassify this work as only interested in "environmental" concerns. Rather, this material and rhetorical work addresses broad earth processes and the complex interrelationships between humans, nonhuman animals, and nonhumans, relations that deserve additional attention because they comprise some of the world's most important relations. Instead of separating material and environmental rhetorics as attending to separate disciplinary concerns, Wells et al. see the production and interaction of all matter as a central concern for rhetoric scholars. "Understood as a process," they argue, this research "evokes not so much discrete perceivable physical entities—whether trees, rocks, or subatomic particles—but rather the energies at work in the ongoing constitution of such things."[55] Rhetoric, especially rhetoric understood as an ontologically powerful force, undergirds the relations that constitute what they call "material life." Echoing Bennett's work, Wells et al. argue that "to speak of 'material life,' emphasizes not what enlivens matter, but rather the liveliness of the material—a liveliness which, long familiar in many venues, is newly gaining prominence in Western thought. In embracing such vitality, many have retreated from the term 'environmentalism,' which can tend to position 'the environment' as a fixed and passive container separate from human activity. Instead, environmentalist commitments have been rearticulated with terms like 'vital materialism,' 'postenvironmentalism,' 'ecological politics,' 'cosmopolitics,' and 'ecology without nature.'"[56] This theory building, and the practical fieldwork that has resulted, has re-attuned rhetoric toward the production of earth materials and the persuasive exchanges generated in their continual construction, while eschewing the more traditional connotations of "environmental rhetorics." The formulation of rhetoric that I advocate for here (and detail further in the next chapter) is situated in this same climate, though it nudges rhetoric toward a more direct immersion within the geologic, because such a framework is necessary for tackling the speculative Earth-alterations of geoengineering deliberation.

In suggesting that rhetoric should be concerned with the "geologic," I mean that the field should attend to the earth as an active agent and constituent part of persuasion—an attentiveness that would not be limited to sediment, magma, or the earth's crust, for example, but instead concern the materiality of the earth itself, atmospheric forces included. The earth plays a constant (though not consistent) role in all other persuasive relations; at times, the planet's geologic forces are the most significant forces in ongoing persuasive relations. It might help here to think of an oft-used comparison called the "Cosmic Calendar," an idea made famous by Carl Sagan. Suppose the history of the universe equates to a calendar year; in this metaphor, the big bang happened on January 1, the earth was formed on around September 6 (about 4.4 billion years ago), and biotic life began around September 14. Anatomically modern humans arrived on December 31 at 11:52 p.m. and scientific rationalism occurred in the final second of the year.[57] When the geologic forces that make up the earth operate in such expansive scales of time, it becomes easy to ignore how they change and how they change us. After all, we are only the most recent additions to this cosmic timeline; yet we live in and with the material forces that ground our very being. Recognizing what considerations of the geologic might mean for rhetoric includes putting the earth back into the rhetorical situation. Here, we would not only have sender, receiver, message, medium, agency, affect, "context," and other forces as elements of a rhetorical ecology, but the earth as an active, omnipresent, constantly changing actor. Of course, simply adding the earth back into a diagram of the rhetorical situation is a fairly superficial gesture, and we would not want to mistake a trivial illustration for real relations. A more comprehensive geologic rhetoric would recognize the impact and significance of earth forces in persuasion and interrogate how that geologic persuasion shows up in policy, opinion, and action. Yet, a geologic rhetoric would not be as stable as an environmental rhetoric with clear commitments to what is unassailably "natural."

Unlike most scientific and environmental rhetorics, a geologic rhetoric is based upon unsteady ground—and here the concept of "ground" has two meanings. By ground, I mean both the literal ground under our feet and the volatile forces that undergird all persuasive acts, because persuasion emerges from, and is built upon, an ontology that assumes certain conditions for the earth. Think of it this way: if the earth has provided you with a relatively unchanging set of geologic conditions, you have likely made a whole host of assumptions about the way the world works based on that stability. For many in the West, this stability has been the default case and our assumptions based on that stability have quietly undergirded almost all other relations and actions. Conceptions of what is real and about how reality functions are based upon this geologic and climactic constancy; the earth successfully recedes into the background and does not intrude into everyday life, even as it supports all life. However, this ground has never been truly stable but has generated the appearance of stability for a privileged few. Further, this ground does not a priori shape all human and nonhuman relations but comes into the present as an existing assemblage of forces that some have taken for granted for far too long. We might call this geologic inheritance a *primordial* set of relations—a set of relations that have preceded others. A geologic rhetoric acknowledges these relations and uses them, builds from them, in developing and critiquing persuasion. A geologic rhetoric also acknowledges that while the earth has the persuasive capacity to shape us, we have likewise shaped it, insofar as humans are a force that has made, and continues to make, definitive and lasting alterations to both the literal and ontological ground. Understood only as the study of intentional, symbolic, human-centric argument, rhetoric would be unable to recognize how geologic forces are generated and continue to play substantial, persuasive roles in the world. But, if more attuned toward the geologic, rhetoric could recognize and then work with these elemental forces. As a practice, rhetorical studies need to confront past assumptions built from a legacy of assumed geologic stability they are to confront climate change and its suggested solutions, like geoengineering, in a way that does not mistake discourse about geoengineering for the complex symbolic-material relations that make up our current ecological situation. In particular, rhetoric should be concerned with the large-scale decoupling maneuvers that are being deployed in the advancement of geoengineering proposals—something I address in more detail in chapter 6.

We should also not be so oblivious as to imagine that "bringing the earth back" into rhetorical analysis is a Western invention, or a rehabilitation of a particular kind of Western rhetorics. For numerous Indigenous rhetorical theories and practices, the earth has been there all along. Rhetoricians should not accept that a geologically attuned rhetorical practice can suddenly "rediscover" the importance of earth forces. To do so would be to mistake a geologic rhetoric as a theoretical position that wakes us up from years of geologic neglect—in much the same way that several Anthropocene narratives claim we have suddenly "awoken" from a long European dream to find that the world has changed forever. Kathryn Yusoff describes this version of the Anthropocene "history lesson" as a patronizing suggestion "in which humanity undergoes a sudden realization of the violent and repressive dimensions of coloniality and then, in the eclipse of modernity, puts these violences firmly in the past, while continuing to perpetuate them through an ongoing settler-colonial present."[58] I take seriously the concern that like claims about the Anthropocene, rhetorical study can be critiqued as a white, Western enterprise. When practiced, a geologic rhetoric should not foreground Western rhetorical traditions by default and thus mistake a geologic emphasis as reinscribing Western rhetorics writ large. By refusing to align a geologic rhetoric with a specific ontological grounding, as that grounding can never be stable, I hope to position this approach as amenable to multiple and diverse investigations, of which this book's approach is just one. A practice of geologic rhetoric should not be, as Yusoff describes, a "paternalism that is tied to a redemptive narrative of saving the world from harm on account of others while maintaining the protective thick skin of innocence."[59] These chapters position how we can emphasize the geologic, material, and discursive forms of persuasion and avoid the mistakes of ignoring the foundational lessons of primordial earth conditions, in order to focus on the harmful and potentially exploitative rhetoric about geoengineering.

In this introduction, I have provided context for understanding how rhetoric can approach geoengineering and show why the persuasive forces being generated in favor of geoengineering should be of great concern to us all. In chapter 1, I establish the foundations of a geologic rhetoric. I argue that because of recent climactic stability and the dominance of what Bruno Latour calls the "modern constitution," it has taken the threat of climactic collapse in Western countries for a renewed attention to geologic and climactic issues to emerge. In response, a rhetoric appropriate for the study of these turbulent earth conditions should be born from the knowledge that there is no such thing as pure "nature" and that to continue such claims is to extend Western modernity and a Holocene fallacy. I trace a line of vitalist thought as a means to question divisions between nature and culture and confound these "horizontal" assumptions, identifying a different lens through which we can see the production of nature and culture as an intermingled affair—a natureculture. Adding the inherited relations from geologic actors into conceptions of rhetoric is no straightforward task but doing so allows us to see the many ways in which our reality is (un)grounded by shifting instability. Unfortunately, while the field of research known as the rhetoric of science should be well-positioned to do such work, it retains numerous epistemological foundations that inhibit its acceptance of this approach. We can identify a more suitable path forward through vitalist thought and Karen Barad's theory of agential materialism, which establishes how reality is comprised of discursive-material forces. While a geologic rhetoric is not bound to a Western theoretical tradition, and geologic-oriented rhetorical forces can also be founded upon Indigenous theories and practices, Deleuze and Guattari's articulation of strata helps me make sense of how discursive-material forces create what can be called an ungrounded ontological grounding for a geologic rhetoric, wherein there is no stable footing from where we can reliably theorize. Instead, a geologic rhetoric is a group of ideas about how persuasion sediments, and we can see this emergent rhetorical work in recent articulations of environmental and material rhetorics. This rhetorical approach is crucially important if we are to confront the large-scale impacts of geoengineering deliberation and counter this rhetoric's emphasis on—and enabling of—SPI as the default geoengineering strategy of the future.

In chapter 2, I pick up a portion of this theoretical foundation and put it to use, analyzing how CO_2 operates as a materially persuasive actor in different networks of power. Starting from the premise that the United States is not in complete stasis about the causes of climate change, I argue that there is room for a surprising agreement within this political context for geoengineering. This chapter engages Annemarie Mol's conception of overlapping ontological layers to understand how persuasion for geoengineering is generated from overlapping discursive-material states. Rejecting the claim that multiple discursively constructed perspectives emerge from one static reality, Mol argues that different material performances can result in different states of being for the same object. By looking at how CO_2 has been generated by two different practices—climate science modeling and environmental economics calculations—this chapter describes the discursive-material practices that have led to two currently competing forms of CO_2. Climate scientists have generated a CO_2 through climate simulations, while economists have created a CO_2 in economic models. At issue is not just linguistic disagreement, or claims about ideology, but the dissimilar formations of CO_2 that have been enacted, over time, as part of a performed ontological reality. Critically, I argue that these potentially competing CO_2 formations could soon overlap in areas where geoengineering policy alliances are being built. This ontological layering could generate persuasive forces that take advantage of scant public knowledge about geoengineering, a lack of international agreements, and an emergent postpolitical state of exception. In this context, climate scientists and conservative think tanks can ostensibly agree about geoengineering research, an agreement buttressed by those who deny that climate change is occurring. Ultimately, competing versions of CO_2 could give rise to a dangerous alignment about SRM strategies in general and SPI in particular.

Chapter 3 considers the danger outlined in chapter 2 and evaluates how formulations of geoengineering have largely been supported by simplified conceptions of risk—understood as a cost-benefit analysis that renders the earth into an economic valuation. Though many geoengineering proposals are purportedly built upon a scientifically defined risk, all compose risk through economic and material conditions. This chapter explains how risk has been determined in geoengineering proposals, and in contrast to earlier rhetorical or discursive constructions of risk, considers the persuasive role of historical, material, and geologic factors in risk analysis. Here, a geologic rhetoric recognizes the origins of climate change risk in a history of Western fossil-fuel extraction and monitors how the climactic effects of that misuse confront us as persuasive relations today. To that end, Ulrich Beck's theory of a risk society is especially useful. Beck argues that we are now experiencing a "reflexive modernity" that must react to these previous geologic abuses. In his view, modern life is dominated by the constant evaluation of risk, in a context where risks like climate change are diffuse and overflow state-level institutional capacities to handle their impacts. Taking up the reigns of control in this reflexive modernity are subpolitical actors (like private industry) who function below state-level capacities for response. In geoengineering, these subpolitical actors have emerged as a "geoclique," scientists and/or entrepreneurs who have coordinated climate modeling projects, influential reports, conference statements, and real-world geoengineering tests. This chapter considers how this geoclique has aligned SPI geoengineering to "natural" analogs, emphasized its potential for private investment, and defined its risk in simulation studies. These formulations of risk threaten to further establish SPI as the default option in geoengineering deliberations.

Asking "what does modern geoengineering research look like?", chapter 4 explores the unusual world of how university research efforts have morphed into current-day geoengineering technology companies. In these two contexts, we find very different discursive-material enactment practices and very different methods of enquiry and production, otherwise known as techne. Studying the world of "ocean fertilization," where iron filings are poured into the ocean as a means of increasing algae growth and encouraging CO_2 absorption, we find that university research-level techne on geoengineering follows strict protocols and makes cautious claims about the data researchers gather. In comparison, companies modeled after technology startups frequently employ a technological solutionism that ignores the carefully gathered data from years of science and sees the thorny problem of climate change through a determinist framework. Often approached through the lens of a single discipline, these geoengineering strategies can fall victim to a form on techne aligned with a "how-to" guide or handbook, or a rational ability to affect a useful result—a techne unsuited to understand complex, distributed phenomena. Encouraged by the tenets of geologic rhetoric to look back on the history of this approach, we see the roots of geoengineering in midcentury technological solutionism and in cybernetics-influenced counterculture environmentalism. What would eventually emerge as ecomodernism began life as an outgrowth of vitalist philosophy and suggestions about how technology could solve the world's largest social and environmental problems. In order to support these claims, I consider ocean fertilization research methods during the 2000s and the eventual for-profit ventures that sought to capitalize on this research, concentrating on how a company called Planktos performed a massive ocean fertilization "experiment" off the coast of Canada in 2012. A geologic rhetoric shows us how for-profit alterations of scientific research either take advantage of claims about what is "natural" or deny the distinction between nature and culture, depending upon which approach best suits entrepreneurial goals. Planktos used both arguments in exploiting a relationship with the Haida First Nation and in using their funds to support this geoengineering project. This experiment provides a distressing example of how capitalism and geoengineering could work hand-in-hand in the future.

For its part, chapter 5 addresses how articulations of the Anthropocene make assumptions about human capacity mirrored in research on geoengineering. Numerous papers on the Anthropocene reference the "noosphere," an idea also cited by the main influences on cybernetics theory. The noosphere is a metaphorical space, unlike the stratosphere or troposphere, and made up of the sum of human cognitive power instead of atmospheric pressure or wind currents. Early thinkers on the Anthropocene connected the noosphere with claims about humans as homo faber, the toolmaking and tool-using species responsible for the coming ecological catastrophe (and the only species that can prevent it). Here, geologic rhetoric helps us locate the species-thinking assumptions made by geoengineering researchers and Anthropocene commentators—who are often one and the same. Species-thinking considers humans as an undifferentiated mass and fails to acknowledge power disparities throughout history or recognize that Western culture has been dominantly responsible for the geologic causes of climate change. Yet, this species-thinking also brings with it assumptions about the human capacity for resilience, and while resilience can be a very good quality, certain definitions of resilience could emphasize SPI geoengineering yet again. Engineering definitions of resilience favor the time it takes to return to a previous global temperature, and this approach could rule out more environmentally preferable resilience strategies. Visualizations of climate change, including some of the best-known ones, can reinforce this form of resilience, further contributing to arguments for SPI geoengineering.

Considering what was learned from previous chapters, chapter 6 addresses the larger forms of harm that could arise from geoengineering and whether the open discussion of geoengineering could present what is called a "moral hazard." A moral hazard is the idea that if geoengineering is publicly presented as a viable option, other carbon mitigation strategies will lose momentum and people will accept a technological solution to "fix" global warming. While researchers disagree as to the severity of the moral hazard question, the issue arises in most interpretations of geoengineering ethics. Even if there is a moral hazard (and I argue there is), the larger problem is the application of paternalistic Western environmental ideologies upon those who will be most impacted by the effects of climate change. Those who advocate for an interventional control argue that Western countries should act practically and decisively to help those disproportionately affected by climate change, and therefore offer utilitarian rationales for geoengineering. Others, like Kathryn Yusoff, explain that Western advocacy for geoengineering risks extending what she calls "White Geology," the continuation of racist policies that extract value from people and geologies to power the global North. Further, any Western-dominant geoengineering policy represents an attempt to rhetorically decouple history from a future where the historical effects of imperialism are incorporated under a supposedly objective execution of a geoengineering technology. Many of the arguments for geoengineering presented throughout this book could in fact be considered decoupling maneuvers, and this chapter showcases how such a rhetorical strategy could create an even greater moral hazard.

Finally, in the conclusion I try to make sense of our deliberative moment and the ethical foundations of focusing on utilitarian, paternalistic ideals for geoengineering. Though I find Western environmental beliefs both comforting and frustrating in equal measure, they are ill-suited for understanding our current epoch. Horizontalist approaches and performative environmentalism fail to scale up to the geoengineering debate and conceptions of oikos are likewise unsuitable for these deliberations. In the context of what Geoff Mann and Joel Wainwright call the "Climate Leviathan," a state of sovereign power structured by environmental management, Naomi Klein's articulation of "Blockadia" is an appropriate response. Blockadia names a range of actions, each grounded in a place, and each responsible to the local, historical, and geologic conditions of that place. Blockadia thus forms the practical action that can be taken in a new geologic hexis, a rhetorical term describing the continually produced state of "having" that we experience in our Anthropocene relations. The choice of which form of hexis we engage in—one guided by economic forces and quick fixes, or one attentive to sustained, careful attention to those geologic relations and supported by environmental justice principles—will determine how geoengineering deliberations play out in the coming years.

In each chapter, I tackle an aspect of geoengineering deliberation that has gone largely unexamined and show a how a geologic rhetoric can be put into analytic and productive use. While this book engages an ostentatious environmental remediation method that will play a significant role in our climate future, it also attempts to bring the earth back into the rhetorical conversation—in a way that previous rhetorical studies have not. My hope is that by giving elements of "the geologic" a place at the table, we can better understand the roles that the earth, and all earth systems, will continue to play in our lives. I hope that in scrutinizing geoengineering deliberations, we advance the conversations, studies, and policies that will continue to form the future of our planet.

1

A GEOLOGIC RHETORIC

One of the most common arguments offered by geoengineering proponents is that geoengineering schemes should be thought of as "natural." Advocates claim that geoengineering can be made to work alongside existing climactic forces and roughly approximated to past geologic events, suggesting that the intentional manipulation of the climate should not be seen as a dramatic, human-instigated technological interference of the climate system but simply an enhancement of that which already exists. Supporters of this view often reference geoengineering's environmental analogs, which is one reason why the SRM "Pinatubo option" comes up so often. Volcanoes are naturally occurring geologic events, albeit often unwanted ones, so any geoengineering scheme that seeks to encourage the same must also be natural. Other geoengineering ideas have also been inspired by similar claims of "naturalness." Strategies that aim to raise the number and volume of clouds by spraying particulate seawater from the ocean into the air are motivated by observing clouds' reflective coefficient. And if you prefer a different rationale, advocates have argued that large ocean-going vessels currently create "ship tracks" (much like those from an airplane), which have already amplified the number of marine clouds. As discussed in relation to the RRAP actions to preserve the Great Barrier Reef, growing the number of marine clouds is often touted as about as benign a geoengineering strategy as possible, perhaps rivalling only the mass-scale heterogenous replanting of the world's forests.

Geoengineering strategies with these supposedly natural analogs have been taken more seriously than before, and studies on geoengineering discourse have indeed shown that appeals to nature can make them appear less dangerous. In their analysis of geoengineering propositions, Adam Corner and Nick Pidgeon note "multiple examples of geoengineering technologies being described by analogy to natural processes and concepts (in both popular and academic publications)."[1] Their work showcases how geoengineering proponents align SPI strategies with volcanic eruptions and carbon capture and storage plans with "artificial trees." The metaphors used to frame geoengineering technologies could also allow for interpretive flexibility and increased acceptance. By connecting their implementation with natural processes, the authors of geoengineering proposals "may wish to make the technology sound less risky than it might be."[2] If we are only "correcting" existing earth processes, this argument goes, then any miscalculations in implementation is unlikely to be catastrophic. After all, they could have happened anyway, so we should presumably have few objections to their application. Yet, the perception of "naturalness," as Luokkanen et al. warn, "may come with a license for free experimenting."[3] This false "natural" perspective is deceptive and ignorant of the latent harm that geoengineering, widely deployed, could cause. And yet, public understanding and acceptance of geoengineering is much like popular representations of nature—a product of discursive and material formations. What we have come to understand as "nature" is constructed both out of what we read and watch and what we experience and engage with over time. While many people might react negatively to pumping sulfate particles into the atmosphere or sequestering excess CO_2 under the oceans, we might also find it difficult to distinguish between what is natural and what is unnatural with any real precision, and therefore hesitate to confidently criticize plans that have been painted as such. In the Anthropocene, what is unnatural about geoengineering? How do we determine, evaluate, and critique geoengineering deliberations?

In this chapter, I argue that the unique features of geoengineering deliberations, including their reliance upon or rejection of definitions of "nature," their dependence upon a kairotic response to an impending environmental disaster, and their promise to shape the entire earth ecosystem, mean that a different rhetorical approach is needed if we are to accurately study and critique these endeavors. They call for a geologic rhetoric, a perspective on rhetoric that brings geologic actors back into persuasion writ large. This integration, however, should not be understood as either a naive addition or one that ignores theoretical traditions from outside Western critical theory; instead, it should be seen as a lens that questions how we ground our assumptions about nature, culture, and the persuasive forces that emerge from within these spaces. This chapter seeks to expose several notions about Western humanities research, and the unstated geologic assumptions upon which this work is often grounded, before questioning why the field of rhetoric known as the rhetoric of science may not be best positioned to tackle these queries. A geologic rhetoric assumes that we have inherited a set of geologic relations that will always be prior to other relations—not as an a priori, but as primordial. Traditionally, a rhetoric of science holds epistemic values that disallows such an approach. Here, I work to make these claims salient and articulate an unstable grounding for a material, environmental, and geologic rhetoric appropriate for the study of world-altering onto-scientific proposals like geoengineering.

A HOLOCENE FALLACY

Conversations about whether a premeditated intervention into the earth's geologic system could be considered natural or unnatural are entwined with preexisting beliefs about the distinction between nature and culture, otherwise known as physis and nomos. These discussions have been a touchstone of Western environmental theory and part of the popular conversation in the United States at least since Marilyn Strathern refused the distinction in 1980. Strathern argued that "there is no such thing as nature or culture, each is a highly relativized concept whose ultimate significance must ultimately be derived from its place within a specific metaphysics."[4] Ultimately, it may matter very much how we frame the environment in relation to geoengineering, to borrow George Lakoff's phrasing, and articulating our current era as the Anthropocene may only further muddy perspectives about potential geoengineering schemes.[5] After all, the Anthropocene is named after the claim that humans have permanently impacted the earth to a significant enough extent that our geologic age can be called, literally, the age of humans. Human influence can be seen in the geologic record, and if we perceive the earth as forever changed by humans and their attendant technologies, geoengineering could slot into our existing worldview as an acceptably natural extension of human activity. Erle Ellis, an environmental scientist affiliated with the ecomodernist Breakthrough Institute has argued just that point, explaining that "the terrestrial biosphere is now predominantly anthropogenic, fundamentally distinct from the wild biosphere of the Holocene and before. From a philosophical point of view, nature is now human nature; there is no more wild nature to be found, just ecosystems in different states of human interaction, differing in wildness and humanness."[6] Numerous Western philosophers and scientists have followed this line of thinking—that there has never been a discrete difference between "nature" and the products of "culture" and that the overwhelming physical evidence for the Anthropocene as a human-impacted geologic epoch only further blurs this binary.

While humanities and social sciences researchers have not adopted the concept of the Anthropocene uncritically, it has sparked a renewed awareness in these academic areas that geologic earth processes play a significant role in human activity. Evidence for the Anthropocene, including plastics pollution, fertilizer misuse, pesticide overuse, and chemical fracking, are now recognized as affecting everyday life. Microplastics are found in almost all species, fertilizer runoff contaminates drinking water with nitrates, pesticides have been toxic to nontarget populations, and fracking can lead to high methane levels in well water. In response, we buy stainless steel straws, practice permaculture methods in our gardens, buy organic vegetables when we can, and vote against politicians who take money from domestic natural gas industries (well, some of us do). Discussions over the Anthropocene have also prompted an awareness that many different humanities fields have ignored the geologic for quite some time. While often concerned with the "environment," the humanities have not often considered the geologic foundations that make environmental conditions possible. Until recently, few humanities scholars had tackled the importance of the geologic in their research—which is not to say that they ignored the environment. Rather, they have focused on the cultural side to the distinction between nature and culture, leaving nature, or physis, to other domains of research. For example, we have trusted physical, biological, ocean, and atmospheric sciences to make definitive, certain claims about the earth. Humanities scholars leave declarations about the natural world to these subject matter experts—and do so with good reason. We have few tools able to prod the chemical compositions, physical reactions, or complex relations of the earth. But the legacy of this disinterest, as Nigel Clark explains, has left "something of a void when it comes to addressing recent scientific claims that humankind has itself inadvertently become a geologic force. Likewise, it [has left] social scientists with few reference points for engaging with questions of intentional human geoclimatic agency."[7] One explanation for this neglect has been the extraordinarily stable climate we have experienced throughout the tail end of the Holocene.

If we base assessments on Antarctic records, "the Holocene appears by far the longest warm stable period" over the last four hundred thousand years.[8] While there have been climactic and geologic changes over that same time period, they have not been as dramatic as the more recent deviations witnessed in the Anthropocene. Clark points out that if we look to the period of time previous to the Holocene, "it's like dropping off a precipice. Between 12,800 and 11,500 years ago, in the midst of a trend towards interglacial warmth, there was a sudden return to ice age conditions, known to paleoclimatologists as the Younger Dryas event. . . . Go back a few thousand years more and this sort of wildly oscillating climate appears as the norm rather than the exception—for hundreds of thousands, perhaps millions, of years."[9] Holocene geologic stability gave rise to an enormous level of human production and creativity, and nearly all the central events of human cultural history (both good and bad) have played out during this timeframe. The livestock domestication, the largescale shift to an agricultural economy, the invention of writing, the emergence of major world religions, widespread urbanization, the emergence of empirical science, and the development of industry, among a host of other expansions, have taken place entirely within the Holocene.[10] Clark contends that this uniquely stable geology, one that has fostered most human achievement, is an anomaly. He quotes anthropologist Brian Fagan's belief in that "civilization rose during a remarkable long summer" and considers the Holocene historically astonishing precisely because of its capacity to provide consistent conditions for complex mammalian life.[11] This exceptional period of stability is both the keystone reason for the emergence of human culture and (much more recently) why Western humanities disciplines have been able to ignore the impact of geologic and climactic processes for so long.

While the humanities have laid claim to greater metaphysical and philosophical truths, albeit less frequently after the emergence of postmodernity, humanities research and theory (rhetoric included) has concerned itself with domains outside the geologic status of the physical earth, objective or otherwise. Within this separation of academic attention, the earth has fallen under the microscope of natural sciences, and in this split, "a division of authority occurred: the history of the Earth and life was the province of natural scientists, and the history of 'progress in human affairs' that of historians and social sciences—a division that still casts a shadow today."[12] The development of the scientific method, the rise of empirical research and the process of induction, and the wholesale displacement of Aristotelian science ushered in a new era in knowledge-making practices. From the seventeenth century onward, the natural world was identified and defined by those outside of the humanities and social sciences, whose interests shifted elsewhere. For Noel Castree and Thomas MacMillan, those aspects of life which were "deemed nonsocial have long been the subject of 'natural science' research and teaching, while putatively nonnatural entities are the preserve of the 'social sciences.'"[13] Carolyn Merchant's canonical *The Death of Nature* outlines the formation of this paradigm.[14] Arguing that Baconian science represented a methodology for the manipulation of nature, she portrays how a mechanistic ontological worldview instantiated a lasting perspective on the framing of the natural world. Merchant contends that the "fundamental social and intellectual problem for the seventeenth century was the problem of order"—and the problem of order was aligned with the messy, confusing, and dangerous "organic" metaphor for life on Earth.[15] The sciences regimented disorder through a mechanistic metaphor that could be supported by empirical study and rational philosophy. Merchant clarifies the extent of this change in Western European thinking, writing, "Rational control over nature, society, and the self was achieved by redefining reality itself through the new machine metaphor. . . . The removal of animistic, organic assumptions about the cosmos constituted the death of nature—the most far-reaching effect of the Scientific Revolution. Because nature was now viewed as a system of dead, inert particles moved by external, rather than inherent forces, the mechanical framework could legitimate the manipulation of nature."[16] F. A. Hayek has described the logical extension of the "removal of animistic, organic assumptions" as "scientism"—the proposition that scientific logic overreaches into areas where it may not be warranted or applicable.

Scientism has been named and articulated by numerous critics, from Hayek to Karl Popper to Donna Haraway, as the hubristic, and often cosmetic, overextension of empirical scientific methods to domains where those methods fail to fit. As Loraine Datson and Peter Galison explain, the late 1800s was a time of faith in "mechanical objectivity," a term they define as "a set of procedures that would, as it were, move nature to the page through a strict protocol, if not automatically."[17] The move toward scientific objectivity reached a peak with the rise of structural objectivity, a twentieth-century ontology defined by "pure" relations, before being called into question by a renewed series of philosophical critiques of science led by Popper, Thomas Kuhn, and Paul Feyerabend, among others. Donna Haraway articulates both the lure of positivist science and its colonizing application as the "search for translation, convertibility, mobility of meanings and universality—which I call reductionism only when one language (guess whose?) must be enforced as the standard for all the translations and conversions."[18] The success of empirical science has led some to think that there are no real limits to scientific methodologies, that there is nothing outside of the domain of science, nor anything to which science cannot by applied.[19] This overreach of scientific thinking has reasserted and reinforced the widespread belief throughout the Western world that nature and culture—the domains of the natural world and the domains of human thought—have always been separate and immutable.

Stuck in this vicious cycle, humanity's perception of its dominance over nature both produces the means to objectify the natural world and then reinforces its separation from culture. In this ontology, nature is an object to be studied; producing a more definitive articulation of the control of nature asserts the distinction between what is "out there" and what are issues of subjective human fabrication and performance. Human activity, technological production, and material creation objectify nature, build epistemic knowledge domains from that objectification, and then reproduce and project the binaries generated from that systemic classification onto all domains of life and nonlife. Such worldviews have also been successfully condemned by scholars in anthropology, ecocriticism, environmental philosophy, feminist science studies, and science and technology studies (STS). Christophe Bonneuil and Jean-Baptiste Fressoz succinctly summarize these objections, arguing that "the natural sciences postulated a physical continuity between humans and other entities that obscured the social production and social relations of nature."[20] Extending this worldview to its impact upon the production of reality, Peter Sloterdijk has called this formation a kind of "backstage ontology"—the often-unstated ontological grounding for empirical knowledge generation and belief systems about the world.

An unstated ordering system grounds beliefs, actions, discourses, and worlds, functioning as an ontological grounding for other theories, worldviews, and philosophies. Bruno Latour has named the assumption of this backstage ontology part of what he calls the "modern constitution," though he also denies that modernity was ever a defensible, cogent description of reality. Latour's modern constitution names historical divisions of power, ones that ordered the world by inventing "a separation between scientific power charged with representing things and the political power charged with representing subjects."[21] To explicate this formation, which had to have been written, constructed, and enacted in order to become an assumed grounding belief structure, Latour looks to the natural philosopher Robert Boyle and the political philosopher Thomas Hobbes. Reconstructing the disagreement between Boyle's mechanistic worldview based in empirical science and Hobbes's defense of the body politic and political theory, Latour highlights how Boyle's and Hobbes's larger disagreements ended up creating simultaneous, but separate, scientific and political discourses. In midst of their ongoing dispute, Latour argues, they invented "our modern world, a world in which the representation of things through the intermediary of the laboratory is forever dissociated from the representation of citizens through the intermediary of social contract."[22] Boyle and other natural scientists built a program that positions the natural world as definable and discoverable in the laboratory, and Latour fills in the details of science's argument, that "nature has always existed and has always already been there; we are only discovering its secrets."[23] Latour goes on to expose the world-building logic underlying the construction of modernity, or as he explains, how science is able to continue constructing nature all the while claiming that the project of empirical science is only discovering what there is to be found.[24] The modern constitution established "a complete separation between the natural world (constructed, nevertheless, by man) and the social world (sustained, nevertheless, by things)."[25]

Latour showcases the power of this formation while also revealing its workings, though he misses out on identifying perhaps the most pernicious force responsible for generating modernity. Walter Mignolo picks up this oversight and has argued for a causal connection complementary to Latour's, but with an important adjustment. *Of course we have never been modern*, because scientific, rational modernity was a project of colonization and racist domination; modernity was *applied to* disempowered and Indigenous peoples as a method of ideological governance. Mignolo observes how the modern constitution was fashioned from Western ontologies and intentionally spread throughout (what is often referred to as) the global South. He considers how the "phenomenon that Western Christians described as 'nature' existed in contradistinction to 'culture' [and how] it was conceived as something outside the human subject."[26] In a contrasting example, Mignolo observes how the Aymaras and Quechuas had no separate concept for nature but instead "saw themselves in it, not separated from it. As such, culture was nature and nature was (and is) culture."[27] He presents this explicit construction of and application of power as part of the broader Western project of modernity, where colonization, the Industrial Revolution, and claims toward rational science ushered in beliefs about the domination of nature. This distinction, and its power imbalances, provided a justification for colonialism, cultural genocide, the wholesale extirpation of Indigenous knowledges, the expansion of slavery and waged labor, and the commodification of biologic and geologic resources. So successful was the backstage ontology of Western modernity that it fueled hundreds of years of coloniality, where nature and culture are key elements to what Mignolo calls a "complex system of Western cosmology, structured theologically and secularly; it also manufactured an epistemological system that legitimized its uses of 'nature' to generate massive quantities of 'produce,' first, and massive quantities of 'natural resources' after the Industrial Revolution. . . . The industrial and the technological revolution also made possible the industrialization of 'produce' and the mercantilization of food and life!"[28] Both Latour and Mignolo describe this ontology as a *cosmology*, emphasizing its construction and performance, and emphasizing similarities with Indigenous worldviews. That said, Latour's proposed resolution to this manufactured cosmology is not to recognize hundreds of years of enforced colonial Western subjugation but to reveal that the strange divisions in the modern constitution could never hold, while noting that we live in a world perpetuated by hybrids of nature and culture—a claim that underscores the pernicious success of the Western ontological project and shows how a stable Holocene, the widespread flourishing of modernity, and its requisite divisions of academic attention have worked to reinforce and sustain a version of reality. Ultimately, climactic and geologic stability assisted in the proliferation of a modern constitution and reinforced Western divisions between nature and culture. If Western geologic worlds were less stable, then perhaps the academic disciplines the proliferated in their wake would have resisted a modern constitution and continued to incorporate geologic considerations. However, the production of modernity created a subtle bias—that of a "horizontalism" that arose as part of the humanities' inability to theorize from the geologic grounding of its own existence.

Our longstanding Holocene stability has created what is known as the "critical zone" of the planet earth. This zone provides for a backstage ontology and likewise generates "horizontalist" thinking. First named by Gail Ashley, a sedimentologist, the critical zone is the surface area of the terrestrial earth defined by the National Research Council as "the heterogeneous, near surface environment in which complex interactions involving rock, soil, water, air and living organisms regulate the natural habitat and determine availability of life sustaining resources."[29] The critical zone is home to almost all conditions for life. Writing with Latour, Alexandra Arènes and Jérôme Gaillardet describe how the critical zone's thin, relatively porous layer has been the sole location for human existence, where "life has modified the cycles of matter by activating or catalyzing physical and chemical reactions."[30] These supportive chemical reactions generate "a kind of skin, a varnish, a biofilm" that has played host to all disciplinary inquiries of the modern constitution.[31] Arènes, Latour, and Gaillardet demonstrate how, through its layers of solar energy, this zone provides the conditions for life, but also how the continued sustenance of this life is startlingly precarious. Simply put, by paying close attention to the significance of the critical zone, we recognize a disquieting asymmetry at work. This thin film supports nearly all human and nonhuman relations and has played host to all human civilization, yet at the same time, this horizontal veneer is precariously supported by a massive volume of vertical geologic and atmospheric actors, often well-outside our field of perception or recognition. The impact that "deep-seated forces of the earth can leave on social worlds is out of all proportion to the power of social actors to legislate over the lithosphere."[32] These forces, relatively stable throughout the Holocene, have resulted in a large-scale obliviousness of their powers, and a concordant ignorance of their verticality, even as human technologies have reconfigured the makeup of this same verticality.

Regardless of our alterations of the earth under our feet and the sky above our heads, the backstage ontology of the Holocene world has been a "horizontal" one where the verticality of the earth has been overlooked in favor of a flattened perspective of human and nonhuman relations. In a horizontalist perspective, we see a limited view of the critical zone, where subterranean and atmospheric conditions are less considered, less valued, and largely ignored by researchers in the humanities and social sciences. By subscribing to this unconscious horizontalism, we disregard, at our peril, the conditions in the atmosphere and the constantly shifting earth. Indeed, we commit a Holocene fallacy whenever we ignore the impact of geologic verticality upon persuasion—a holdover of thinking from the Holocene epoch. A Holocene fallacy is the belief that the earth is relatively stable and will be for the foreseeable future; because of this assumption, a Holocene fallacy also occurs when we fail to consider the deep history of the earth or the many ways that vertical conditions impact our lives in almost incalculable ways. Most Western humanities research commits some version of this Holocene fallacy, including all but a few studies in the field of rhetoric.

Like Mignolo, sociologists Paul Carter and Heidi Scott align these horizontalist conventions with modernity and colonial imperialist politics. Carter argued that the entire colonial era was dominated by Western "scenographic conceptions" of space that imagined the world as a "continuous planar surface," regardless of Western powers' ambitions to mine, destroy, and devour their colonized landscapes.[33] Concerning these ambitions, Scott claims that we need to begin constructing an alternate politics of the subterranean, in part because the "undeniable 'horizontality' of colonial and imperial expansion should not be allowed to overshadow the dimension of verticality that, in many times and places, was equally central to colonial and imperial ventures and was manifest in practices concerned with the subterranean such as mining and the mapping of geological formations."[34] The long, stable Holocene not only contributed to the modern constitution, corralling and restricting disciplinary knowledge—it also generated a controlling Western horizontalist representation of existence within the narrow band of the critical zone. With the collapse of the Holocene, a humanities research paradigm built upon the assumptions of a stable geology should similarly be exposed as unstable. Investigations that seek to alter earth systems critical for sustaining life, like geoengineering, put pressure on longstanding horizontalist perspectives, because they require an intense awareness of geologic and atmospheric verticality. While a Holocene humanities was built upon a stable geology, a geologically attuned Anthropocene humanities can no longer make these same suppositions. Any inquiry into the techno-environmental practices of geoengineering employing a Holocene perspective would likely replicate horizontalist, Western, modern claims and therefore fail to approach these concepts from the geologically accurate context in which they were formed. Accordingly, our inquiry must begin with different assumptions.

THE GEOLOGIC HUMANITIES

With the erosion of our Holocene stability and the anthropogenic destruction of our fragile and necessary critical zone, geologic conditions are now asserting themselves as a persuasive force. The term some artists have chosen to represent this moment, the "geologic now," signifies a growing awareness that these unstable geologic actors are here to stay and that the modern constitution's hard distinctions between nature and culture have never been fully viable. The "geologic now" names the significance of the geologic, where the term "geologic" does not reference the field of geology so much as it alludes to a range of earth-centric actors, forces, and influences, as well as to earth systems and processes. Focusing on the temporal dimensions of this shift, Elisabeth Ellsworth and Jaime Kruse have used the term to describe how artists and designers have been exploring the significance of the geologic in all aspects of life. They claim that "until recently, the word "geologic" conjured meanings and associations that referred simply and directly to the science of geology. . . . But that seems to be changing. . . . recent events and developments are making geologic realities sense-able with new physical intensity and from new angles of thought."[35] The "geologic now" names the state in which these insistent Earth instabilities have presented themselves. These forceful actors include global warming, surface radiation, forest fires, plastic deposits, ocean acidification, new erosion and runoff patterns, cataclysmic weather events, rising sea levels, melting glacial and polar ice, fracking earthquakes, oil pipeline spills, and a host of others. It is through these actors that the geologic is being emphasized, and while modernity, as an assumption of geologic stability and a horizontal worldview, has historically shielded many in the West from this instability, these volatilities are becoming increasingly felt by all.

To better understand the logics and possible practices within the geologic now, the humanities have been forced to become more attuned to nonhuman, earth-centric actors and actions, incorporating those features into theoretical and methodological positions. We might argue that the humanities are undergoing a kind of disciplinary "geologic selfwitnessing," what Kathryn Yusoff defines as "the praxis of a *geo-logic* of life forces as geologic forces."[36] Geologic selfwitnessing is an awakening, an attentiveness that humans are "an end in themselves—both as a distinguishable trace in the geologic strata and as author of the end—a stratigraphic writer, reader, and agent of geology and nature."[37] In rhetorical studies, we can trace similar forms of selfwitnessing through work done in the name of rhetorical ecologies. Nathan Stormer and Bridie McGreavy's work on the significance of ecology to rhetoric, for example, reads ecological thought as an organizing ontology that helps reframe rhetorical theory.[38] Inspired by studies in aquaculture, they borrow ontological commitments in ecology to rethink the commonplaces of rhetoric through capacity, vulnerability, and resilience, focusing on the material relations these commonplaces can be used to explore. Their work, and the work of others on rhetorical ecologies, like that of Candice Rai and Caroline Gottschalk Druschke and Amy Propen, can be said to be primarily concerned with the geologic now.[39] A more complete understanding of our ongoing geologic moment requires additional collaboration and cross-disciplinary borrowing, such as Stormer and McGreavy's incorporation of ecological thought or Druschke's suggestion that rhetoric can be understood through "the flows of relationality between beings," where rhetoric literally is the energy transfers between constellations of life and nonlife in ecologies.[40] These investigations are clearly connected to the emergence of the geologic now, though no work in rhetoric has approached geoengineering as of yet.

Typically, when tackling a topic like geoengineering, the humanities are put under tremendous pressure to understand and engage a world that has been largely absent from previous research. Some of this pressure comes from scholars themselves, because while we believe that those in the geologic, atmospheric, and ecological sciences can bring vast amounts of insight to bear on the state of the Anthropocene, we also suspect that "geologists, meteorologists and their colleagues are not necessarily well-equipped to study the sort of things that take place between humans . . . the composition of a rock or the pattern of a jet stream being rather different from such phenomena as world-views, property and power."[41] Geoengineering is ultimately a human-centered technological and scientific proposal and often fits squarely within existing environmental perspectives that center human need and intention, like preservationist or conservationist environmentalism (albeit often with an ecomodernist spin). The humanities are especially well equipped to deal with these kinds of human relations, even if we need help understanding other causal relationships along the way, and even if we have yet to explore the geologic. Christophe Bonneuil details how fields like environmental history, natural anthropology, public policy, human ecology, sociology, political ecology, and ecological economics, among others, have begun to engage in more complete dialogue with the natural sciences, in hopes of generating additional knowledge about the Anthropocene and its challenges.[42] Not included on Bonneuil's list are humanities disciplines like literary studies, new media, and, most importantly for our purposes, rhetoric.

Literary critics have been incorporating geologic concepts into their critical apparatuses, exploring literary production and meaning through the concept of "planetarity." Originally defined by Gayatri Chakravorty Spivak in *Death of a Discipline*, planetarity refers to the relationship between the human and nonhuman world and functions as Spivak's way to rethink global relationships in the context of aggressive global capitalism.[43] In asking us to "imagine ourselves as planetary subjects rather than global agents," Spivak argues that we can begin to see ourselves not as humans in control of an abstract Earth but as beings that must necessarily live on and cope with a physical, omnipresent planet.[44] Christian Moraru has further extended planetarity in *Reading for the Planet: Toward a Geomethodology* and with Amy J. Elias in *The Planetary Turn*. Moraru articulates planetarity as the world's condition or modality of being, an ontological shift of discursive-material production that marks "the world's reemergence qua planet, in early twenty-first-century literary, cultural, and theoretical practices."[45] Spivak, Moraru, Elias, and others attend to the political ideological implications for literary study, though it is clear that a broader acknowledgment of the importance of the earth's history is now emergent in other humanities disciplines, too.

For example, scholars in media studies have also focused on geologic history, with some attending to nonlinear media history, such as Siegfried Zielinski, and others to the material and geological "deep time" of media forms themselves, such as Jussi Parikka.[46] In *A Geology of Media*, Parikka argues that geology can help address the underlying physical state of media, where the "metal, chemical, and mineral agents of history become pushed from mere contextual insights to agents of a different sort of genealogy."[47] His emphasis allows for a new attention to materiality, one that concentrates on "a different sort of temporal and spatial materialism of media culture than the one that focuses solely on machines or even networks of technologies as nonhuman agencies."[48] For Parikka, this work is needed in an Anthropocene epoch where capitalism extracts as much energy as possible from all of the actors it touches, human and nonhumans included. Parikka wants the humanities to recognize the importance of "biological and geological contributions as part of the social collective," as part of a larger shift whereby we can see humans as geological agents resisting the Cartesian dualisms that continue to influence disciplinary thinking.[49] His work aligns with both horizontalist critiques and the concerns of the geologic, as Parikka explains that geology is "not only about the soil, the crust, the layers that give our feet a ground on which to stumble: geology is also a theme connected to the climate change as well as the political economy of industrial and postindustrial production. It connects to the wider geophysical life worlds that support the organic life as much as the technological worlds of transmission, calculation, and storage. . . . It becomes a conceptual trajectory, a creative intervention to the cultural history of the contemporary."[50] Instead of the study of media by means of the messages it is able to carry, Parikka is interested in the material formation of media itself—how digital messages are still indebted to the "mineral durations" that are essential to any communication.

Likewise interested in how the geologic past has shaped and continues to shape culture is the interdisciplinary area of "energy humanities," or "petroculture" research and criticism. While we could point to older works like Debeir, Deléage, and Hémery's *In the Servitude of Power: Energy and Civilization through the Ages* as speaking to similar themes and ideas concerning energy production and its influence on power and culture, more recent petroculture research could be centered around Canadian cultural critic Imre Szeman.[51] Szeman's writing, including *After Oil*, *Energy Culture*, and *On Petrocultures*, as well as several edited collections, describes the commitments of the field.[52] In *Petrocultures: Oil, Politics, Culture*, Szeman and his coeditors, Sheena Wilson and Adam Carlson, note that the humanities have finally become aware of the extent to which all modernity is indebted to oil and energy production and is awash in the significance of fossil-fuel-based kilocalories.[53] "At the heart of this newfound awareness," they argue, "is our recognition that over the course of our current century we will need to extract ourselves from our dependence on oil and make the transition to new energy sources and new ways of living."[54] Through their work, Szeman and others have explored the depth of modernity's debt to energy extraction and production and the ideological, political, and colonial enterprises that have supported this energy generation. Szeman and Dominic Boyer explain that petrocultures research tries to "shed light on the fuel apparatus of modernity, which is all too often invisible or subterranean, but which pumps and seeps into the groundwaters of politics, culture, institutions, and knowledge in unexpected ways."[55] Yet, because critique is never sufficient, researchers in this field also attempt to create the "critical intellectual resources by which we might make possible the social transition needed to support and enable energy transition."[56] In some sense, this means that petroculture studies are speculative but focused on the deliberative rhetorics of what could be accomplished with useful ideological and practical grounding.

All of this research represents how current humanities research can interrogate ecological and material concerns and how these recent efforts are part of a much longer, more subtle history of criticism that supports an explicitly geologic rhetoric. We can connect these theories as lying under the broader heading of what William Connolly calls the "eccentric" or "minor" tradition of thinkers. He groups together theorists such as Heraclitus, Friedrich Nietzsche, Mary Shelley, Bruno Latour, James Baldwin, Donna Haraway, Gilles Deleuze, Eduardo Viveiros de Castro, and Karen Barad—theorists "who insist upon the periodic volatility of nonhuman processes, who emphasize the importance of nature/culture imbrications of diverse sorts, who invest many nonhuman actants with purpose and striving, who often enough acknowledge tragic possibility in the world, and who often embrace the sweetness of life in a world not highly predisposed to organic belonging, capitalist mastery, or communist control over nature."[57] Connolly claims this line of thought contrasts with the humanities in general. For him, most Western humanist scholars have been "rampant enablers, long too uncritical of extractive practices and the pursuit of fossil extraction, high productivity, and consumer abundance in capitalist, socialist, and communist societies."[58] Instead, he argues that theorists of this minor tradition—which I would argue could also include Spivak, Moraru, Elias, Parikka, and Szeman—"are entangled with multiple and heterogeneous processes and seek to appreciate and cultivate those entanglements."[59] The trajectory of this thought breaks from mainstream Western humanities work that has ignored the geologic, because its practitioners include complex earth processes within and throughout their theorizing, not just as a "topic" for analysis or an applied ethics.

Most humanities research, rhetoric included, has not accepted the ideas, inferences, and claims made by this minor tradition. As evidence for this lack of attention, we only need to look at how the earth's critical zone has remained largely unnoticed by Western scholars—that is, until the stakes and conditions of its grounding began to shift more rapidly than before, thus exposing its fragile verticality. Such vertical "undermining" not only upsets assumptions of a stable geology but also throws the very conditions of stability into a new light, showcasing how geologic actors support work done in such fragile environs. These revelations are now being interrogated in Anthropocene theory and prompt us to ask questions that have previously been the domain of geology, climatology, and other earth sciences. With the recognition that we have never been modern, that we are always geologically from within a critical zone of relations, that we have deep-seated horizontalist assumptions, and that those assumptions are always buttressed by powerful vertical actors from above and below, we can confront geologic relations with fields and methods previously untested.

A rhetorical approach to the geologic must necessarily note the conditions where persuasion has been grounded, state the circumstances of that grounding, and recognize the ongoing destabilization of that grounding throughout time. That "grounding" includes the assertion that a stable critical zone tacitly supports and is subliminally built into rhetorical theory. It can be as easy to identify as the implied assumptions of predictable, seasonal change, for example, or the awareness that certain chemical preconditions for life support the persuasive relations that follow—or it can be as complex as the historical, ontological emergence of molecules in industrial chemistry and the profound effects they have upon different environmental and ideological commitments. It should also prompt us to interrogate how the rhetorical strategies used by others rely on particular geologic assumptions. Because geologic conditions in the critical zone play such a substantial role in the formation of persuasive relations, we should grant those conditions a similarly outsized function in how we understand persuasion. Persuasive forces are supported with, surrounded by, and comprised of geologic conditions that function in a way that could be called primordial (more on this term shortly). When used to analyze existing persuasive situations, like geoengineering, a geologically oriented rhetoric should seek out how geologic conditions influence the discursive and material formations that occur. As Nigel Clark argues, investigations into the geologic may "lead us back to epochs before humans emerged, take us deep into micro-ecologies too tiny and too multitudinous to even imagine, drag us down to the molten and lifeless interior of the earth, [and] whip us up into the stratosphere and out into the solar system."[60] Intentional, human persuasion has been situated upon the earth, wholly within the critical zone, but it also makes use of the earth in ways that only an analysis of the geologic preconditions of persuasion can expose. Further, because all persuasive relations leverage the geologic in the generation of power—some more than others—it is also the role of rhetorical analysis to render discernable those intentional and unintentional uses of the geologic. In later chapters, we will see how some arguments for geoengineering attempt to ignore or manipulate these primordial relations, and how doing so is also an ethical, geologic, and rhetorical failing. Throughout this book I attempt to show how geoengineering proponents mobilize the geologic in different ways in their pursuit of research opportunities, economic advantages, or political policies, but this study is simply one means to see these commitments through. Other geologically attuned rhetorical studies could tackle a range of concerns and topics that have traditionally been considered the domain of environmental rhetoric.

For the purposes of this investigation, however, a geologic rhetoric helps interrogate the destabilizing, world-changing relations suggested in deliberations about how to alter the climactic conditions of the earth. A rhetorical theory appropriate for the study of something as geologically intense as geoengineering should mirror Yusoff's work on world-altering ontology. She argues that any form of geoengineering involves the

> reconstitution of the earth as a dynamic world object. This is not just the imagination of earth-as-artefact, as in the Apollo images from space or the future-orientated earth objects of climate modelling (Edwards, 2010), but it involves an understanding of the earth as a geoengine that can be altered, modified, and engineered on a global scale. The engineering, then, is not metaphorical, so much as bio-geophysical; it involves the conceptualisation of an abstract, nonlinear biodynamic model of the earth in which atmospheres and earth systems, and even the sun, are imagined as "engines" (albeit organic) open to forms of making. This is not about creating new worlds per se (although geoengineers do now talk about "ideal climates"), but about a shift from the earth as a given to the earth as characterised by ideal forms of output, such as climate.[61]

Studying geoengineering deliberation thus requires different ontological suppositions than previous regimes of rhetorical theory and should take into consideration what Yusoff calls the "broader marketisation and management of geophysical and biochemical flows in ecosystems and the atmosphere."[62] Rhetoric, in other words, should attend to the onto-geologic relations inherent in persuasion. The implications of this change for the field of rhetoric are therefore a fundamental concern, one foundationally different than that composed in previous studies of scientific rhetoric. Yet, existing assumptions made in the area of rhetorical study seemingly best-equipped to tackle geoengineering appear to be ill-suited for this kind of work.

RETHINKING THE RHETORIC OF SCIENCE

The rhetorical approach that would seem most appropriate to the study of such a vast and complex scientific project like geoengineering is the rhetoric of science (RoS). However, in reality RoS (as it has been articulated until very recently) is a less-than-applicable methodology here, due to some of the field's longstanding conventions that run against the attention to the geologic grounding described earlier in this chapter. For much of its history, scholars in RoS have studied scientific texts, practice, and policy through the lens of rhetorical analysis, often employing concepts from classical rhetoric, communication theory, and critical theory.[63] RoS is currently a healthy field, involving active participation by scholars in the Association for the Rhetoric of Science, Technology, and Medicine, conferences aligned with the Rhetoric Society of America (RSA) and the National Communication Association, publications in technical, scientific, and environmental communication journals, and dozens of monographs and edited collections since the 1990s. According to Jeanne Fahnestock, the goal of a rhetoric of science is to "understand the means, the how of texts, movements, or argument fields that recruit adherents for their claims about the natural world. To explain the workings of the available "means" of persuasion in any particular case."[64] The field has covered a great deal of ground pursuing the available means of scientific persuasion, including studies of major scientists and their writing, the genres of scientific writing, instances of scientific controversy, the use of inventional practices, the communication and definition of risk, the rhetoric of medicine, and the formation of scientific visualizations, among other lines of inquiry.[65] However, traditional RoS approaches would need to change their tenets to usefully examine something as in-flux and world-changing as geoengineering deliberation. For example, it will come as no surprise that RoS is first-and-foremost committed to understanding how scientists make use of language in generating disciplinary knowledge. Scholars have concentrated on "argument and audience; on tracking tropes, figures, and commonplaces; and the careful examination of the production and reception of argument."[66] RoS scholars deem rhetoric to be the intentional, human use of symbolic language, commitments largely due to the field's historical theoretical positioning.

While RoS can point to a number of locations for its forbearers, most would, like Alan Gross, aim their literature reviews at Aristotle. Gross identifies both the early sophists and Aristotle's *Rhetoric* as starting points for a rhetoric of science, even though Aristotle would have largely disagreed with the premise of the field.[67] That is, Aristotle claimed that science was that which could be verified, whereas rhetoric's domain was only that which was variable—that which could be otherwise. RoS also owes its theoretical conventions to Chaïm Perelman and Lucie Olbrechts-Tyteca's *The New Rhetoric*, the work of Stephen Toulmin, and the strong program in the Sociology of Scientific Knowledge.[68] Research in RoS was built upon concepts developed in the sociology of science that were themselves built from claims in the philosophy of science.[69] According to Steve Fuller, "as the story is normally told, rhetoric was the direct beneficiary of inroads that sociologists made into philosophical accounts of science."[70] Likewise, in his articulation of the distribution of rhetoric of science scholars, Ken Baake contends that "the rhetoric of science program . . . is indebted to sociologists and philosophers of science who helped lift the veil of objective infallibility from science."[71] Inheriting a critical stance from Kuhn and Feyerabend, as well as W. V. Quine, sociologists argued that science was an epistemic project that could be understood through sociologic methods. This paradigm gave way to claims about the social construction of science, thereby providing "the overwhelming bulk of the warrants for social constructivism, a Protagorean epistemology whose strongest credentials come from analyses of scientific discourse."[72] Major works from this time include Barry Barnes's *Scientific Knowledge and Sociological Theory* and David Bloor's *Knowledge and Social Imagery*.[73] Barnes and Bloor pursued an "epistemic relativism" or "factual relativism," stances which placed scientific knowledge as something socially constituted, open for debate, and deeply dependent upon historical, cultural, and local factors, including the discourse and culture of scientific disciplines themselves. With these assumptions, science can then be interpreted as "relative" to any number of external, local conditions, which suggests that scientific knowledge, instead of providing a one-to-one correspondence to the truth (and mirroring "scientism") is simply one of many possible conclusions to which researchers could have arrived. Additional work on the history of scientific discovery linked scientific truth with historical and cultural attitudes, and specific studies of scientists and their laboratories connected claims to localized discourse communities. Often called the "strong program," or the Edinburgh School, this research tried to avoid positivist claims of scientific realism and instead framed scientific truths as contingent.

Emergent from this sociological tradition, RoS often adopts a postmodern rhetorical stance that argues, in its "strongest" instantiations, that symbolic language can fully account for the truth claims that scientists have generated, though not for reality itself. While few RoS researchers claim such a radically discursive position, many would agree that the field's primary concern should be the symbolic representation of science, where such representations convey how disciplines generate knowledge about the world. In essence, we could say that RoS has been primarily concerned with epistemology: how symbolic communication plays a role in determining truth claims. Few RoS scholars would disagree that rhetoric is epistemic, and several have announced this position loud and clear, including Alan Gross, in a foundational statement.[74] Emergent as part of the larger "rhetoric of inquiry" movement, RoS engages how scientific knowledge is made and how language, visuals, and other symbol systems generate and sustain meaning across time. Here, rhetoric is defined as the "intentional use of persuasive language."[75] Rhetoric of science scholars, whether they come from communication or English departments, often begin with these assumptions.

The rhetoric of science has been grounded within a horizontalist framework, one that is interested in the science taking place in the critical zones of human habitation but has so far been neither willing to recognize the limited verticality of that work nor especially concerned with the ontological conditions that make it possible. There is little possibility in this research for an altered ground that emphasizes a geologic primacy for the conditions for rhetoric. That said, as of this writing RoS also appears to be changing tack somewhat, moving away from just the epistemic study of science as it has been represented in symbolic language or visuals. In considering the future for RoS, Carl Herndl and Lauren Cutlip argue that the field "is shifting from a modern and humanist disciplinary focus to a non-modern and post-human focus. This shift includes a change in our dominant theory of realism, our understanding of agency, and the location where we do our work."[76] These changes can accommodate the disciplinary assumptions of a "geologic now" and allow for rhetorical study to focus on the molten interior of the earth, the dizzying heights of the atmosphere, or the chemical bonds that make it all possible. While not all members of the RoS community are willing to give up the methods that have given rise to an independent area of study, many are clearly aware of the growing split in the field. A good representation of this division could be seen at the 2018 RSA conference, where the term "rhetoric of science" was worried over in several presentations. At least three panel discussions focused on this rift, and several self-consciously placed researchers with different methodological and theoretical alliances in dialogue in order to either challenge or defend the assumptions in RoS research. David Gruber and Randy Allen Harris performed one of these confrontations, with Harris explaining how "RoS is about scientific texts (broadly construed; I'm including beakers here, and colliders)," and Gruber countering that "RoS can no longer be limited to being a field content to investigate scientific texts, big T or little t."[77] These panels included queries about the future of RoS as a research area, concerns about the field's increasingly strained relationship to rhetorical criticism, and hypotheses about what happens when long-standing tenets in the rhetoric of science are transgressed.

If we were to approach geoengineering from more traditional RoS assumptions, we could perform an analysis of the documents generated by the climate scientists, conservative think tanks, technology start-ups, and universities that have contributed to its development, using those documents as representations of scientific knowledge-building practices. That work would undoubtedly be useful; however, to understand the persuasive forces that have been operating and continue to operate upon geoengineering deliberations, rhetoric cannot simply study the texts of scientists but should instead be attuned to the geo-ontologic relations that Yusoff signals, including how the power of geologic actors is being marshaled into persuasive material relations about the earth's future. This rhetoric should likewise not uphold distinctions between nature and culture, because to do so would project an epistemic belief onto a science that no longer recognizes the validity of such a framework and because geoengineering research now largely operates within Anthropocene, ecomodernist assumptions. In order to carry out this analysis, we require a new grounding for rhetorical theory. By this I do not mean a "grounded" theory approach, but a theory that emphatically maintains a geologic asymmetry in tackling the human-nonhuman persuasive relations that proliferate in the Anthropocene.

A RHETORICAL COUNTERNARRATIVE

A rhetorical approach appropriate for studying the coproduction of matter in the Anthropocene should avoid the modern distinctions between nature and culture. That much is clear. We would not want a geologic rhetoric to concede claims about what is real to scientists; instead, it should argue that things in the world are coproduced in moments of symbolic and material becoming, where nature is not the opposite of culture but created along with culture in a dynamic and recursive process. Further, we need this work to recognize how verticality is ontologically foundational in persuasive relations and frame this recognition in alignment with what I have referred to here as the "geologic." One way to rethink the epistemic and representational commitments of RoS investigations is through a counter-Cartesian rhetorical theory that tackles more than just knowledge-generation practices. In *Ambient Rhetoric: The Attunements of Rhetorical Being*, Thomas Rickert laments the position of rhetorical theory as only engaging in symbolic representation, noting that "rhetorical theory has yet to find a way to come to terms with symbolic mediation without occluding the material world except by recourse to standard theories of realism."[78] Still, while his work offers one path out from this bind—through an ambient rhetorics—the most applicable model for a rhetorical reimagining of the rhetoric of science might come from Byron Hawk's work on vitalism, in his counterhistory of rhetoric and composition. Rhetorical vitalism and its contiguous theoretical concepts offer much for countering the strong divisions between symbolic, human-centric knowledge production and naive materialism.

In *A Counter-History of Composition: Toward Methodologies of Complexity*, Hawk reintegrates the forgotten history of vitalism—and its oft-maligned ontology—into the history of rhetoric and composition, proposing that vitalism is better attuned to explain nuances of contemporary rhetorical contexts.[79] Vitalism, in brief, holds that the impetus for reality cannot be explained through a mechanistic worldview and that an inherent energy is integral to the emergence of all life—and depending upon the flavor of vitalism, all nonlife, too. Hawk reconsiders vitalism and its place in rhetoric in order to link eighteenth-century German organic philosophy to Samuel Taylor Coleridge's philosophical vitalism, as part of his larger claim concerning how a number of German Nature-philosophers sought out vital concepts that would unite subject and object (and reject a division proposed by Immanuel Kant and German idealists).

Vitalist theories in general are suited to geologic rhetoric because they identify all matter in the process of becoming and recognize that the world is not a fixed entity made up of static forms. We could contrast vitalism, roughly categorized as "process metaphysics," with the more dominant philosophical tradition of "substance metaphysics." Substance metaphysics holds that things *are what they are* and can be traced back to Parmenides's Eleatic school of thought. Parmenides's poem *On Nature* largely covers the pre-Socratic philosopher's perspective on reality, which is divided into things that could not be otherwise and things that are variable, and generally showcases his perspective on the unchanging, indestructible material state of nature.[80] We see this same delineation later in Greek thought and beyond, notably in Aristotle, who confines rhetoric to study areas that could be otherwise—proofs that are variable and not simply "found" in nature. Process metaphysics opposes this tradition with a different set of assumptions, namely by replacing the description of static forms with new categories that define things in the world by *what they do* and *how they occur* in time. Things in the world are therefore not stagnant entities ready to be catalogued but dynamic formations replete with a long history of their own emergence. One philosopher who reenergized up this line of vitalism was Friedrich Wilhelm Joseph von Schelling, an "oppositional" vitalist whose work offers a different location from which a geologically attuned rhetoric could be established. Schelling's vitalist process metaphysics is especially useful for rethinking rhetoric in the Anthropocene, in part because he offers us a different foundation from which we can deconstruct the nature/culture binary and determine matter in all its complex relations.

Schelling's work confronted how human cognition and the energy for the production of life function together. For him, nature was not the sum of all physical laws in their current state but an interplay of challenging forces that oppose each other, hence Hawk's framing of Schelling as an "oppositional" vitalist. More specifically, Schelling argued that opposing forces are central to how all matter is produced, contending that "in resolving the problem of how matter in general is originally possible . . . the problem of a possible universe has also been solved."[81] What Schelling meant is that opposing forces create new matter, and he used a metaphor of a stream (perhaps borrowed from Heraclitus) to make this point.[82] Namely, when a stream "meets resistance, there is formed a whirlpool; this whirlpool is nothing fixed, but something that in every moment is vanishing, and in every moment springing up anew."[83] He assumed that the world is always in the process of becoming, and not made up of fixed, stable entities best understood through substance metaphysics. Schelling saw forces of nature and forces of thought as "phases of the same ground," and according to Hawk, "argued that both science and philosophy should work to show how matter starts as force and through dynamic processes becomes mind."[84] Hawk recognizes how Schelling made nature "a purposive system with a *telos* [where] the world becomes a dynamic process of movement from object to subject caused by the tension between two polar forces of the same power."[85] Because oppositional forces generate the symbolic and material production of matter, Schelling was also able to claim that there is no difference between cognition and nature, or between what we would consider nomos and physis. Diana Coole, commenting on Schelling's work, claims that he made "no essential difference between organic and inorganic nature . . . they are merely potencies with different powers of organization such that inanimate matter becomes living being through its internal development."[86] While Schelling later turned his attention to art and aesthetic production, and other vitalist philosophers moved away from an insistence on oppositional forces, Schelling's ideas influenced other sciences for years to come, including the then-emergent field of ecology and specifically one of the originators of Western ecological science, Alexander von Humboldt. Schelling's writings prompted Humboldt to write *Kosmos*, one of the most popular nineteenth-century books and arguably the cornerstone of botanical geography.

Schelling suggested that all matter is always in the process of being produced, and while we must necessarily engage with the material constraints of those productions, we can also engage *in their production*. This vitalism "sees function and internal relations as a founding principle rather than as external, visual characteristics."[87] What Hawk explains as "internal relations" and "vitalistic and mechanistic theories" are the beginnings of what would come to be called cybernetics theory, a worldview that plays a substantial role in the development of geoengineering throughout the twentieth and twenty-first centuries. Henri Bergson, who popularized vitalism in the twentieth century, positioned these relations against the forms of order in the universe, where mathematical order was aligned with a positivist science (and what we would call scientism). Opposed to this system, Bergson described a kind of vital force, an élan vital of the driving power that promotes dynamic and internal growth, adaptation, and change. This framework is essential for numerous conceptions that are still relevant to philosophical debate today, including Bergson's defense of the emergence of homo faber (the toolmaking and tool-using human) and his explanation for the diversity of life on earth. These ideas are also useful for understanding geoengineering discourse, as is Bergson's provocative suggestion that matter—all matter—is always involved in a dynamic process of becoming.

Bergson claimed that to understand how a thing is comprised, we must look at the history of that material's emergence—an insight that could have been pulled straight from Schelling's work. He proposed that "matter, looked at as an undivided whole, must be a flux rather than a thing," which suggests that mechanistic and vitalist forces are constantly present.[88] Such a dynamic view leads to the conclusion that any investigations of matter and its relations should interrogate how it changes over time. Further, it implies that science alone cannot explain these histories and forces. Bergson contended that even though it seems as if matter is defined by scientific epistemologies, materials "endure *only because* they are bound up inseparably with the rest of the universe."[89] Iris van der Tuin has argued that Bergson's take on the production of matter is surprisingly analogous to modern articulations of the discursive-material production in new materialist thought.[90] She reads Bergson's work as a precursor to Karen Barad's articulation of "agential materialism" because their remarkably similar concepts of material intra-action allow for unique agents and agencies to emerge. In *Meeting the Universe Halfway* (2007), Barad argues the same point.[91]

Specifically, Barad contends that matter never only refers to a stable entity but is instead a dynamic formation of relations. Matter, she claims, "does not refer to a fixed substance; rather, matter is substance in its intra-active becoming—not a thing, but a doing, a congealing of agency. Matter is a stabilizing and destabilizing process of iterative intra-activity."[92] What Barad describes as "matter," or "phenomena in their ongoing materialization," can also be explained as material agencies in the process of exerting their powers.[93] In calling for an understanding of scientific or technical practices as the performance of commingling agencies and forces and not just the creation of a thing or the discovery of a natural property, Barad argues against positivist formations of science, which propose that "the apparatuses that produce phenomena [can] be understood as observational devices or mere laboratory instruments."[94] Scientific instruments alone are not clear indicators of reality, but instead devices intimately connected with practices for generating symbolic meaning through material relations. When practicing science or engineering, she claims, we engage in "making specific worldly configurations—not in the sense of making them up ex nihilo, or out of language, beliefs, or ideas, but in the sense of materially engaging as part of the world in giving it specific material form."[95] Conversely, Barad also refuses the claims of postmodern discursive constructionists, because symbolic meaning is not the final stop in generating reality, where "language does not represent states of affairs, and measurements do not represent measurement-independent states of being."[96] Social-constructivist accounts that assume a representational stance fail to take into consideration the agential capacities and powers of nonhumans. For Barad, at issue is not "some ill-defined process by which human-based linguistic practices (materially supported in some unspecified way) manage to produce substantive bodies/bodily substances but rather a material dynamics of intra-activity: material apparatuses produce material phenomena through specific causal intra-actions, where 'material' is always already discursive-material—that is what it means to matter."[97] By explicating that knowledge practices have material sources, Barad indicates that all performances have specific material engagements with the world.

In tracing this trajectory from Schelling to Barad, we identify a path through Western critical thought that rehabilitates material relations as fundamental to ontological conditions, incorporates claims about reality, responds to the nature/culture divide, and disallows both positivist science and social construction accounts of knowledge. Here, persuasion occurs through the ongoing discursive and material formation of reality, and the practice of rhetorical analysis should be able to study these relations for the acts of science that emerge. A geologically attuned, discursive-material rhetoric asks us to contend with the ontological formations that have coalesced into the present and empowers us to not make horizontalist assumptions or commit Holocene fallacies. These are some of the conditions to which a geologic rhetoric must be adjusted if it is to be attentive to Anthropocene geoengineering. Yet, we would be mistaken if we pursued a critical path that is ignorant of, or wholly independent from, other critical traditions that have made similar claims.

Western critical theory has often been oblivious to the connections between new materialism and Indigenous theories of materiality, where nonnaive realism emerged out of Western scholars "searching for a theory and practice of inquiry that preserves the incisive examination of our practices of knowing characteristic of the best poststructuralist social analysis, but that retains an ability to acknowledge the protean materiality of things as a source of meaning"—a line of inquiry that Schelling, Bergson, Latour, Barad, and as we will see, Deleuze and Guattari, also attempt.[98] Rosiek, Snyder, and Pratt have explicitly called out Western critical traditions for their lack of engagement with Indigenous conceptualizations of nonhuman agency and human-environmental relations. Their work bridges this gap by locating Barad's claims alongside those of Indigenous theorists, seeking productive points of interrelation. "In Indigenous studies scholarship," they argue, "non-human agency is taken as a given and so is less frequently introduced as a general concept. Instead, there is more emphasis on the formation of relations with particular other-than-human agents."[99] For many Indigenous theorists, "there are repeated and consistent references to an understanding of the character of agency as something that emerges out of particular circumstances in such a way that its most salient features are missed if it is dealt with primarily as a general abstraction."[100] Some of these relations with nonhuman agents are assumed in land-based ontologies, as those worldviews are often grounded in how reality is manifest in specific physical locations. In her work on land-based literacies, Gabriela Raquel Ríos positions Indigenous relationality as assuming that "humans and the environment are in a relationship that is co-constituted and not just interdependent," where relations are made with specific sites and specific lands.[101] Rosiek et al. criticize Western new materialist theories for "re-discovering" material ontologies and for being largely ignorant of Indigenous conceptions of human-nonhuman relations. They echo Kim TallBear, who argues that "the invisibility of our ontologies, the very few references to them in their [Western] writing, and reference to indigenous thought by other theoretical traditions" further negates Indigenous ontologies in new materialist theorizing.[102] Too often, when they are referenced by Western research, Indigenous ontologies and claims about materiality are deployed in ways that alter their original meaning and reinforce a false distinction between nature and culture.

Vanessa Watts's work on Western theory-building recognizes similarities and differences between that work and Indigenous place-thought. Watts claims that Western theorists often use origin stories as examples of "symbolic interconnectedness—an abstraction of a moral code. . . . often distilled to simply that—words, principles, morals to imagine the world and imagine ourselves in the world."[103] Counter to this interpretation, she finds that in many Indigenous origins, humans arrive on the planet *after other humans and nonhumans, and after what we might call geologic events*, where "the inclusion of humans into this [existing] society meant that certain agreements, arrangements, etc. had to be made with the animal world, plant world, sky world, mineral world and other non-human species. Therefore, being associated with animals, whether it be through clan systems, ceremonies, or beings that acted as advisors, transpired from a place of reverence."[104] Watts explains how Indigenous ontologies have long been framed by Western worldviews, highlights the damaging dismissals of Indigenous human-nonhuman relations, and names the different groundings upon which Indigenous ontologies are positioned. It would be challenging, as Watts notes, for Western ontologies to accept some of the essentializing categories of different Indigenous cosmologies, due to political and theoretical differences in each tradition. This difficulty is also evident in theorist Mel Chen's hesitancy to venture outside of "secular" theorizing in engaging with Indigenous ontological traditions, as noted by Kim TallBear and others.[105] Chen describes this hesitancy as guided by Trinh T. Minh-ha's ethics of not speaking about, or speaking for, but rather speaking nearby other ontologies.[106] I share this concern, and the project I outline here as a means to tackle geoengineering and rethink the rhetoric of science does not abstract an Indigenous ontology for the purposes of theory building, as that would likely constitute an appropriation, but instead emphasizes that Indigenous theorizing of material relations and place-thought ontologies offer a long-standing, different, and (hopefully) coordinative theoretical stance that can complement the geologically attuned rhetoric I locate here. One parallel example could be Robin Wall Kimmerer's articulation of Ojibwemowin grammatical ontology, where land features like bays and hills are understood as "to bay" and "to hill," thereby refiguring a falsely assumed stability as an ongoing material performance.[107] This ontology is not unlike Barad's attention to the practice of producing and structuring the significance of matter. An Indigenous theoretical tradition should be recognized alongside the minor Western trajectory from Schelling to Barad outlined earlier in this chapter, in overcoming Cartesian, modernist assumptions and in providing alternative ontological grounds.

Within this complex of new materialist claims, Indigenous ontologies very much included, we see nature, physis, as more than a location—and more than an inspirational space—but constituted in ongoing relations. Equally, we are able to render the geologic as a co-constitutive, where matter is emergent through symbolic and material interaction. The emergence of a "geologic now" is therefore an overdue recognition of the importance of the geologic in Western thought, one that is necessary to tackle world-making practices like geoengineering, and one that has the additional capacity to engage Indigenous ontologies and what Heidi Scott has called the "(post)colonial subterranean" as means to explore Western expansionary adventures.[108] Theoretical trajectories like Connolly's minor theorists, Barad's agential realism, and Indigenous ontological thought can be used to break down philosophical distinctions between earth processes and human processes—a methodology that can express a geologic rhetoric and apply that framing as it pertains to complex discursive-material assemblages. What remains, then, is to state the conditions of this new grounding for geologic rhetoric and its application.

UNGROUNDED GROUND

Understanding the Anthropocene in terms of Barad's articulation of a discursive-material world-making casts "nature" as a tragically mutable configuration. Humans have succeeded in shaping the earth to such an extent that our influence is now part of the geologic record. However, our intentional and unintentional alterations do not end there. As a result of our relations with various nonhuman geologic actors, our current political, ethical, and philosophical crises are permeated by ecological, atmospheric, and geologic volatility. Timothy Morton asks what will become of the figurative ground (of theory) when the literal ground upon which we build our claims becomes unstable. "The more we become aware of the dangers of ecological instability," Morton argues, "the more we find ourselves lacking a reference point. . . . we're losing the ground under our feet at the exact same time as we're figuring out just how dependent upon that very ground we are."[109] While the naming of the Anthropocene has become a kind of "wake-up call" about the significance of geologic for Western scholars, this awareness is also deeply unsettling, because it has revealed itself only through its changeability.

Kathryn Yusoff's work once again locates a crucial distinction for the significance of the geologic. For her, the naming of the Anthropocene provides a break with an older formation of geologic ground. It has offered "an aesthetic-political concept for that realization of geologic life, allowing for possibilities of thinking the earth as a generator of those modalities; it names this sensation of becoming-geologic; the rupture of the end of the Holocene and climate stability; and a new state of geophysical earth-becoming that promises to be less forgiving in the *removal of the ground, both literally and figuratively*."[110] What is this "removed" ground upon which we are living? Perhaps obviously, the ground has been recognized to be more *literally* unstable. While geologic processes have occurred for billions of years, and the earth has been known to be mutable, human-caused geologic instability has obviously increased. Mudslides caused by deforestation, earthquakes caused by fracking, wildfires caused by poor forest management, rising sea levels caused by increasing global temperatures, ocean acidification caused by amplified atmospheric CO_2, and other instabilities are altering our world. These new revisions, which we can witness not only in our lifetime but often in everyday life, are evidence enough that the supposedly solid Holocene grounds upon which Western culture's dominant formations have been built are now unsteady. Here, "ground" refers simply to the ground we walk on. For the second meaning of an unstable "ground," Gilles Deleuze and Félix Guattari's work on the status of ontological conditions proves valuable.

In *A Thousand Plateaus: Capitalism and Schizophrenia*, they utilize geologic concepts to explore the ground in both senses explicated here—as literal ground under our collective feet and as a grounding for ontological relations. They set up what they call a strata-analysis as a key methodological gesture. Strata, Deleuze and Guattari explain, are "Layers, Belts. They consist of giving form to matters or imprisoning intensities or locking singularities into systems of resonance and redundancy. . . . strata are acts of capture."[111] Mark Bonta and John Protevi further define strata as "a freezing or congealing of consistencies," where "actual or 'stratified' substances are equilibrium, steady state, or stable systems, and hence display most prominently extensive properties . . . which are by definition divisible without a change in the kind of system . . . and definite qualities."[112] A shorthand (and albeit somewhat reductive) way to think of strata are as congealed assemblages of significant actants.[113] Yet, these strata are not islands onto themselves—they continuously exchange properties and relations with neighboring strata. Stratification is therefore not a process of division but rather one that enables "confrontation with the spatial arrangements of the social divisions of materiality and the claims to power that are enlarged by harnessing the geopower of the substratum."[114] Here we see the discursive-material production of the geologic as similar to Barad's argument for the production of matter. Through discursive and material interrelation, then, strata produce reality, where the social, nomos, is unstably grounded by the geologic.

Deleuze and Guattari's strata, as layers of intensities and energies, are formed through a process with which geologists would be familiar: sedimentation. This sense of sedimentation, however, is somewhat different than the forces that create physical rocks; according to Yusoff, "to sediment means to push hard upon the layer below, to contract and concentrate its forces, and thus to intensify rather than to simply layer upon it."[115] In Deleuze and Guattari's terminology, while there are "hierarchies of order between groupings," there also exists "a succession of framing forms, each of which informs a substance and in turn serves as a substance for another form."[116] The result of this interrelation means that geologic strata are made up of social forces, and social strata are influenced by geologic pressures. While it might be tempting to see this as the social construction of geology, that reading would replicate mistakes made by postmodern social constructionist arguments and misunderstand the connection between nature and culture. Instead, "geological and social planes," Yusoff argues, "are not opposed, and work upon one another at a distance and in proximity," resulting in explicitly "geosocial formations."[117] Geosocial formations are where geology and social life interact, and where the more foundational, precedent strata of the geologic provide an unstable set of conditions upon which social life is performed. Geosocial formations are the intermingling of the geologic and the social, and the geologic is therefore more foundational than we might first assume.

As we extract resources from beneath the earth, make planetary decisions, and alter the makeup of the atmosphere, we must necessarily contend with the inherited geologic conditions that ground all life. As Deleuze and Guattari note, each strata formation "serves as a substance for another form."[118] Because the geologic curates an unstable ground for the social, our relationship with the earth is coalescent with the ecological conditions that appear at-hand within the critical zone of our inhabitance, but asymmetrically so because geologically causal strata partially preconfigure our being. These geologic conditions, which we inherit as necessary for existence, ground a significant portion of any ontologizing that takes place. Clark asserts that this geologic grounding "suggests a formative and fundamental asymmetry. However turbulent the ground may be, it implies a form of relating in which some actors rely on the achievements of other actors in ways that exceed any measure of mutuality or co-dependence."[119] The asymmetry of these inherited conditions may appear to be an a priori condition, but a better term might be "primordial," meaning "the origin or starting point from which something else is derived or developed, or on which something else depends."[120] The geologic forms the primordial physis for all nomos, and through their interaction, offers up the unstable grounds for our geosocial reality.[121]

The strata that Deleuze and Guattari articulate thus form the literal and figurative ground, one that is coproduced but never fully controllable through human action alone. They argue that "we inhabit a planet . . . that is liable to spontaneously reorganize its elemental strata and flows into novel configurations."[122] The geologically unstable ground becomes the unstable grounding for all the formations of being that occur on, in, and with the earth. The earth is "far too volatile to ever function as an anchoring ground. Its integrative role is merely an interval before the earth shakes off its composure and reactivates its constitutive elements. . . . Effectively, what Deleuze and Guattari's earth bundles together are all the terrestrial capacities for upheaval and metamorphosis. Their earth is a metastratum of inherently excessive forces and energies, in which even the most apparently stable or stratified formations will sooner or later unsettle themselves."[123] The earth, the geos, *is* the unsteady ground we stand on both literally and figuratively, because it provides for both a physical and metaphysical grounding. That said, no discursive-material formation, no matter how *primordial*, can provide a fully stable ontological footing from which philosophical or rhetorical claims can be made. Yusoff recognizes this as well and explains that there is "nothing deterministic about how matter becomes expressed within social formations, it is not pre-differentiated as such, but it is through the process of interlocking with the differentiated materialities of the earth that these identities become defined (and so social and geologic strata are mutually defining but not commensurable)."[124] With this relationship so articulated, we can draw a connection between Barad's theory of agential realism and Deleuze and Guattari's geologic ontology: for both, symbolic forces and material forces combine to enable new matter into existence in a way that eludes the control suggested by human-dominant formations of agency. However, the distinction between the two, and where Yusoff and others build upon Deleuze and Guattari's work, is in how the geologic simultaneously grounds and ungrounds any ontological formations it contributes toward.

We can take away two major implications from this understanding: (1) the geologic is comprised of social forces and can never be understood as purely, naively material, but rather as a pressurized series of relations, and (2) while the geologic is foundational to the social, and while all social relations must necessarily deal with preexisting geologic ones, those relations are neither fixed nor stable. Deleuze and Guattari succeeded in debunking the illusion of human-controlled agency by showing that social strata do not belong to humans alone and "that what humans do accomplish on this stratum depends on forces that are not necessarily under human control."[125] As Nigel Clark puts it, "In the Deleuzoguattarian universe, any manifestation of matter-energy can bed down into strata. But the earth—this actual planet—has special significance."[126] The forces not in our jurisdiction are most certainly the complex, vertical, unstable, geologic actors that have been made apparent in the Anthropocene. They are forces that "rear up and stretch their pincers out in all directions at all the other strata."[127]

A MATERIAL, ENVIRONMENTAL, AND GEOLOGIC RHETORIC

We can chart a geologic rhetoric approach to geoengineering in several ways, primarily through the recognition that any rhetorical study should include geologic actors, not just as an afterthought but as a deeply foundational part of analysis and production. While suffused with discursive, cultural forces, the geologic underpins our reality through an unstable grounding, and any account of persuasion should necessarily incorporate these relations. When approaching a world-altering scenario like geoengineering, this means that simplified, horizontal accounts of rhetoric fail to acknowledge how a geologic history interacts with the discursive and cultural factors behind specific geoengineering approaches. If we ignore this, we risk applying a theoretical framework from Western modernity onto a world that lives by no such rules. Further, a geologic rhetoric should reveal how geologic strata support our current ontologies as primordial forces—underpinning discursive formations of reality, imbuing scientific rationality with cultural history, and highlighting harmful manifestations of discursive-material production that could promote certain groups or certain technologies. As a broad methodological approach, geologic rhetoric is responsive to recent work in rhetorical ecologies and could manifest in the productive analysis of other scientific deliberations beyond geoengineering.

The significance of assuming an ungrounded ontological ground has also emerged in rhetorical work on the environment, as rhetoricians who study traditionally "environmental" interests also begin to consider how unstable, nonrepresentational, and material relations support complex ecological domains. The aforementioned rhetorical ecological work by Nathan Stormer and Bridie McGreavy pursues a similar purpose, as they pull from ecological thought to reconfigure rhetorical commonplaces. Their efforts are part of a larger redefinition of the nature/culture binary, as means not only to study the environment but also to provide a wholesale reconfiguration of rhetoric in the Anthropocene. Similarly, Nathaniel Rivers has argued that Anthropocene "environmentalism specifically needs a more intense rhetoric—one engaged not simply in human discourse, but in the nonhuman, in the object."[128] Rivers is responding to how rhetorical theory has been marshaled to engage environmental topics, deployed in the writing classroom through ecocomposition pedagogies, or merged with environmental and ecological metaphors to make larger claims about both. Thinking with environmentalists William Cronon and Thomas Birch, Rivers advocates for a "deep ambivalence" to wilderness, insofar as ambivalence assists us in thinking through the more complex material-symbolic relations that exist in the natural and not-so-natural world and keeps us from assuming a stable ontological footing for our claims. Reconsidering the environmental and the rhetorical together, Rivers pursues this ambivalence through object-oriented ontology and new materialism to complicate more traditional Western environmental rhetorics that privilege humans, reinscribe the boundary between people and things, and reinforce humans' supposed "alienation" from nature.

Other rhetorical work in this vein includes studies of Anthropocene environmental conditions and the discursive-material analysis needed in order to engage them. Amy Propen's investigation of the role of images in the visual-material shaping of human-nonhuman relations positions this analysis as not possible in terms of symbolic representation alone, but instead as stories that "must be told in a way that acknowledges and respects the ongoing relationships of all beings, matter, and bodies."[129] This extension of rhetoric is only available through vitalisms that are ultimately supported by Schelling's views on nature and culture, and by vitalisms that highlight how the natural world is emergent from entangled symbolic and material relations. Likewise, rhetorical efforts that perform the kind of research that Rivers suggests here include Jennifer Clary-Lemon's material and participant-engaged *Planting the Anthropocene*. Her "new materialist environmental rhetoric" is also aligned with vitalist thinkers and Connolly's minor tradition of European theory. Her study blends "elements of ambient rhetoric, new materialism and material ecocriticism, and critical affect studies and . . . is resonant with postcolonial and Indigenous research paradigms that value relational ontology."[130] Like other material rhetorics and rhetorical ecologies research, Clary-Lemon's theoretical framing "is both human and nonhuman, one attuned to the movement-oriented persuasion of bodies and things" and the realities of post-nature worlds.[131] In the expanding field of environmental rhetorics, we can see how the discursive-material construction of matter and the primordial status of the geologic are used to understand a more complex rhetoric.

In these new studies of rhetorical relations, the geologic (often framed as ecology, nature, or the environment) is a foundational precedent that is central to any rhetorical formation, yet, frustratingly, we cannot depend on any of these geologic forces to remain stable as we project them onto what we could or should do on the future. A geologically attuned deliberative rhetoric must be constantly attentive to the geologic—as primordial, certainly, but as a primordial ooze—a precedent that will always be changing. Rhetoric, as the critique and performance of relational capacities, is ultimately and unassailably *of the geologic*, even though the geologic can never provide a universally committed set of grounds upon which we can build a theory, perform analysis, or produce discourse. All subjects, not only "environmental" ones, deserve to be studied with these geologic concepts, and the historical co-composition of the physical earth should be a key feature to any rhetorical investigation.

The field of rhetoric can successfully examine world-altering plans like geoengineering, avoiding a simplified horizontalist thinking amplified by the modern constitution, and instead follow how the primordial forces of the geologic exert their influence. This geologic rhetoric can offer much to the study of geoengineering, especially as many plans actively reimagine the future of our earthly habitation. There are numerous geoengineering programs already running, including significant studies backed by the University of Oxford, the University of Calgary, Harvard University, Arizona State University, and the Royal Society, not to mention the many independent governance projects currently in progress. These projects detail what geoengineering could look like, who is allowed to deploy such methods, and when those deployments should take place. These deliberative engagements are also crucial for the future of our planet, as they will play a substantive role in the co-construction of nature, particularly as geoengineering continues to move out of the laboratory and into the wider world. We can trace ongoing deliberative plans for this geologic co-construction through the emergence of climate change science, the effects of industrial modernity, the exploits of rogue geoengineers, and the rationalization of species-thinking. It is these locations that give us insight into what deliberation about geoengineering looks like and why it has taken that shape. Rhetorical study is thus deeply important to understanding geoengineering deliberation; we need to pay very close attention to "how we, collectively and heterogeneously, might negotiate more carefully, more judiciously, more generatively with strata."[132] The following chapters attend to the geologic forces at play in arguments for geoengineering and examine how they are marshaled into power relations that suggest future action. At present, they reveal some of the most worrisome features of this deliberative moment.

2

CO_2 AND STASIS

In 2009, in the middle of a field in rural Russia, a group of climate scientists loaded a huge aerosol generator onto a military truck and proceeded to spray sulfate particulates across the landscape. The researchers watched as the cloud reached a height of about 650 feet and then, having installed a number of sensors at ground level, they measured the temperature of the atmospheric surface layer through the aerosol haze. Though their methods have been described as "extremely subscale" by some, the results published in the journal *Russian Meteorology and Hydrology* led to the authors' conclusion that "it is principally possible to control solar radiation passing through artificially created aerosol formations in the atmosphere with different optical thickness."[1] The study was mostly ignored by major Western news outlets, but this experiment constituted the first field test of a particle injection SRM technique, and perhaps the first explicit field test of a geoengineering concept. The chief scientist behind the test was Yuri Izrael, former vice-chairman of the IPCC and occasional advisor to Vladimir Putin. Izrael opposes the Kyoto Protocol, claiming that it has no scientific basis, and has previously argued that "there is no proven link between human activity and global warming," because "there is no scientifically sound evidence of the negative processes that allegedly begin to take place at such temperatures."[2] Deeply skeptical of anthropogenic global warming, Izrael is conversely optimistic about the future of geoengineering and has suggested that "the technological possibilities of a full-scale use of the method will be studied . . . and optimal latitudinal zones and altitudes of injection into the stratosphere should be estimated to get the maximal effect with minimal costs."[3]

Unfortunately, there has been a growing interest in and acceptance of geoengineering projects like Izrael's—ones that tackle the effects but not the causes of global warming—an interest that has much to do with how CO_2 has been fashioned by different groups. These groups have not formed coordinating realities about CO_2; rather, they have created the conditions for consensus around which some geoengineering strategies will find more acceptance than others. In this chapter, I examine the present geoengineering moment by probing one of the key geologic reasons why these ideas are enjoying increased attention as of late. But before we fully delve into modern-day geoengineering plans and strategies like Izrael's, we need to attend to the vertical sources of climate change—the chemical compounds at the root of the West's current lack of agreement over global warming. A rhetorical study of CO_2 is essential if we are to understand geoengineering because CO_2 functions as more than a static chemical compound, and more than a cultural imaginary; rather, it works as matter whose discursive-material relations have resulted in starkly contrasting versions. Differences about what CO_2 is and how it functions are perhaps the central issues in the lack of agreement about climate change, and therefore the central issues in geoengineering deliberation. From a rhetorical point of view, we can easily say that discussion around climate change has not reached stasis, a state of agreement about the core elements of a discussion. Working with the properties of the compound, climate scientists' modeling efforts have emphasized a CO_2 with specific features, as have influential economists, conservative climate change skeptics, and romantic "green" environmentalists. Each of these groups has played a role in the production of CO_2, as it were, though they fail to agree on what CO_2 is and how it performs in the world. The relations surrounding CO_2 have generated multiple states of being, wherein CO_2 functions in distinct ways but could also result in very different political groups aligning in their commitments to geoengineering. I should be clear: my goal here is not to argue that wildly different forms of the chemical compound CO_2 are floating around the earth; rather, it is to define CO_2 as it has been performed into reality by different groups, not simply through discursive construction but in discursive-material acts that bring different forms of matter into mattering. These versions of CO_2, while different enough to be called incompatible, could create what some have called overlapping "ontological layers" in ways that support consensus, and stasis, on geoengineering. These persuasive alliances benefit the riskier, but supposedly cheaper, SRM geoengineering strategies and suggest that they possess an advantage over slower and safer, but more expensive, CDR technologies.

CLIMATE CHANGE, STASIS, AND ONTOLOGICAL LAYERING

The reasons why we disagree about climate change, and therefore climate change solutions, can be located in a range of different factors: poor science communication, differing psychological responses, social pressures, religious beliefs, political machinations, infrastructural fixity, and the difficult-to-comprehend scale of climactic cause and effect. However, the opinions of those who deny some (or all) of the claims of climate scientists on global warming still register loud and clear in current public policy debates. These "denialist" views include the outright rejection of global warming, the disagreement that greenhouses gases produce increased atmospheric temperatures, and the disavowal that climate change is anthropogenically caused. Leah Ceccarelli has described these kinds of maneuvers as "manufactured scientific uncertainties," borrowing the term "agnogenesis" from Robert Proctor to describe intentionally created feigned ignorance, doubt, or obfuscation.[4] Frustratingly, there is evidence that these strategies have worked. A 2018 study found that while roughly 75 percent of Americans believe global warming is happening and 62 percent "understand that global warming is mostly human-caused," about 23 percent still claim temperature increases are due mostly to natural changes in the environment."[5] While (as of this writing) rates of climate change denial in the United States are the lowest since the Yale Program on Climate Change Communication has been studying the issue, there still exists significant disagreement about whether climate change has been predominantly human caused or not, despite the overwhelming scientific affirmation that it has. The disagreements that we have about climate change are so contentious in part because we cannot seem to settle on some of the most basic elements of the discussion, thanks in part thanks to denialist rhetoric, and our continual disputes mean that resolutions have remained elusive. One valuable way to understand this lack of agreement is through the concept of stasis, which names a state of discourse in which definitions for common terminology have been agreed upon and can be used in productive debate. When a topic has not reached stasis, the parties involved have not agreed to the same definitions for the terms of their discussion, and this lack of equivalence prevents meaningful movements toward resolution. Many of the most complex topics of modern-day American public policy have not reached stasis, including abortion rights, gun control, and climate change. Central to our current lack of stasis about the latter are differing definitions of CO_2.

Stasis is a strategy for rhetorical invention, whereby the process of asking specific, heuristic questions about the nature of the topic being debated generates information about the key issues within that topic. Stasis can be reached, or has a greater potential to be found, when the answers to those questions align. The word "stasis" literally refers to a "stance" and was originally used to describe the position of a boxer toward their opponent.[6] Cicero treated the four major parts of stasis theory (borrowed from Hermagoras) as heuristics applicable to any controversial matter, and his descriptions in *Topica* are perhaps the clearest representations concerning problems of fact ("Is it?"), problems of definition ("What is it?"), problems of nature or quality ("Of what sort is it?"), and problems of action ("What action—if any—is appropriate in the given case?").[7] Stasis theory is especially useful in analyzing legal disputes, but for debates that are more broadly social than those argued in a courtroom, like the public discussion over climate change, considerations of stasis allow us to identify the complexity of ongoing disagreements.

Through stasis theory, we can appreciate how climate change denier discourse is often more nuanced than the mere disbelief of scientific data. Michael Mann identified different steps on what he calls the "ladder of denial" of global warming, roughly corresponding with the different elements of stasis. He points to a "hierarchy" with six stages of denial, wherein "contrarians have tended to retreat up the ladder of denial as the scientific evidence has become more compelling," through a concession of different points:

1. CO_2 is not actually increasing.

2. Even if it is, the increase has no impact on the climate since there is no convincing evidence of warming.

3. Even if there is warming, it is due to natural causes.

4. Even if the warming cannot be explained by natural causes, the human impact is small, and the impact of continued greenhouse gas emissions will be minor.

5. Even if the current and future projected human effects on Earth's climate are not negligible, the changes are generally going to be good for us.

6. Whether or not the changes are going to be good for us, humans are very adept at adapting to changes; besides, it's too late to do anything about it, and/or a technological fix is bound to come along when we really need it.[8]

While the "retreat" to accepting climate science is not as linear as this ordered list might suggest, Mann's categories roughly correspond to Cicero's. Position 1 is an issue of fact, 2 and 3 are related to definition, 4 and 5 concern the impact of that definition, and 6 addresses future action. The primary disagreement over climate change, and likewise the key factor in arguments for geoengineering, is what we might call "causal"—the cause-and-effect relationship between human activity, increased atmospheric CO_2, and increasing temperatures. Therefore, the United States' current lack of stasis is best represented by disagreement over stages 3 and 4 (and possibly 5) of Mann's ladder. Without some basic consensus as to what CO_2 is and how it functions, we might not see any further agreement over how to resolve climate change.

The core of this stalemate involves those who argue that climactic data tells a clear story about how CO_2 operates in the atmosphere and those who argue that these causal connections are unclear, unreliable, or have been manipulated by special interests and/or scientists. But disagreements over what climactic data means and how that data has been used are not only about scientists' motives. For many, CO_2 has also been marshaled into convoluted economic relationships that define how it functions. For climate scientists, temperatures are increasing, an increase that has been caused and is still being caused by human activity—much of it performed knowingly—and has resulted in increased global temperatures and climactic instability (among other effects). Climate scientists consider the chemical and physical attributes of CO_2 when making climate simulations, and those models ask CO_2 to perform in a certain way. But for some economists, increased atmospheric CO_2 *may* cause increased global temperatures, the impact of GHGs on temperature *may* be negligible, and global warming *may* be a natural, cyclical occurrence. For them, CO_2 is defined primarily by its position in a financial nexus. However, if we were to stop here and state that our disagreements about the reality of climate change are rooted in a lack of agreed-upon definitions of CO_2, we would be ignoring the impact of the geologic in this persuasive context. For example, we might be tempted to blame a lack of stasis on the discursive construction of CO_2, or we might fault certain groups for refusing to recognize that CO_2 has chemical properties that impact atmospheric temperatures. The former position ignores the material production of reality and the latter takes materiality as an absolute; neither suggests that reality is itself produced through an intermingling of discursive and material forces, and both disregard how CO_2 has emerged, over time, to function the way that it does today.

A geologic rhetoric highlights the discursive-material production of CO_2, and in doing so, identifies how it manifests in different climate and economic models over time. It is through this approach that we are able to recognize the different relations of CO_2 that have been ordained as part of performed reality. Examining the history of how CO_2 has been generated in different ontologies not only reveals the past discursive-material development of the compound but also highlights its role in generating points of agreement. While different assemblages of people and things have produced incompatible versions of CO_2, there also exist places of overlap between these ontologies—places of overlap that generate the aforementioned state of exception over geoengineering and could advantage certain geoengineering solutions over others.

To have different forms of CO_2 does not mean that there are two different forms of CO_2 floating about; it means that CO_2 has been performed into being in distinctive ways by disciplines, people, instruments, measurements, and associations—and that all those relations play a role in enacting a reality. But to discuss how something can exist in multiple realities, we need to consider how objects like CO_2 are engaged in these performances, practices, and enactments. John Law has defined an "enactment" as the "relations, and so realities and representations of realities (or more generally, absences and presences) . . . being endlessly or chronically brought into being in a continuing process of production and reproduction."[9] An enactment is practiced by people employing tools and methodologies to bring about a repeatable result. Most scientific disciplines can be defined as performing various enactments. Law argues that these practices multiply over time, as there is not just one set of enactments functioning on the materials that make up our world. "The argument," Law states, "is that there isn't one set of practices. Instead there are lots of them. And those practices . . . are all enacting and re-enacting putative realities."[10] A reality is generated by an array of people and things, acting together, collectively enrolling different objects in their relations. The practice of making, experimenting, and simply employing objects in different performances can render them into ontologically coherent states of being—ones that can be incoherent to other materials enacted in different practices. For Law, this is "the third great version of STS, the turn to enactment. We are still in the realm of the real, the ontological."[11] Such an argument might appear like the tired social constructivism of the culture wars but is fundamentally different in scope: acknowledging multiple ontologies shifts the conversation from the discursive or social construction of objects toward the distinct performed realities in which an object is enmeshed. Social constructions leave materials at the command of discourse, but a theory of multiple ontologies recognizes the interrelated discursive-material production of reality.

To understand reality as multiple, and to analyze multiples of things existing in those realities, we also need to emphasize that reality is performed. As Annemarie Mol explains, "rather than being seen by a diversity of watching eyes while itself remaining untouched in the centre, reality is manipulated by means of various tools in the course of a diversity of practices."[12] Objects are not simply produced through language or symbolic actions but are rather performed in different ontological assemblages. Earlier, I argued for a similar distinction in the difference between epistemology and ontology. Similarly, Mol maintains that "epistemology is concerned with reference: it asks whether representations of reality are accurate. But what becomes important if we attend to the way objects are enacted in practices is quite different. Since enactments come in the plural the crucial question to ask about them is how they are coordinated."[13] By foregrounding practice, we can identify how objects are created within and as part of those relations and habits. "And since the object of manipulation tends to differ from one practice to another," Mol clarifies, "reality multiplies."[14] The act of employing objects in practice, which must necessarily be both symbolic and material, can render them into ontologically different states of being for those that make use of those objects. In *Ontological Politics in a Disposable World*, Luigi Pellizzoni also explains how ontological states are formed. Namely, in his view, "traditional accounts of representation are replaced by performative gestures that create the world they hint at, or meld into."[15] Instead of epistemic representation, therefore, we are in the realm of performed ontology.

Bruno Latour has made similar claims about the nature of scientific discoveries and examining his oft-cited discussion of the ontological properties of microbes can help us understand how ontology operates. Notably, Latour attempted to resolve the seemingly simplistic question concerning Louis Pasteur's discovery of microbes. For him, the two commonsensical ways in which we could answer the question "Where were microbes before Pasteur?" are both traps. On one hand, we could say that Pasteur invented or discovered microbes, and that before his discovery, they did not exist. In this answer, Pasteur gets agency over microbes and writes microbes into being (or maybe being into microbes), starting from the moment the discovery was proven in a publicly adjudicated arena. The obverse to this thinking is the claim that microbes were there all along and Pasteur was merely the first human to identify them. Here, the status of microbes as ontologically real is safe and the role of science is simply to continue disclosing our given reality. Seemingly, the two arguments cannot be resolved. However, much of Latour's answer to the conundrum is built upon his claim that any new discovery does not mean either that the phenomenon became real only after a "Eureka!" moment in a laboratory or that it simply existed forever. Latour argues that any existence, microbial or otherwise, requires ontological enactment "in a costly and massive institution which has to be monitored and protected with great care."[16] In this ontology, both humans and nonhumans rely on each other as something emerges into reality, by the force and significance of their performance. Instead of either assuming that Pasteur and the practice of "science" constructed microbes, or that microbes were there all along, Latour argues that while Pasteur generated the conditions to make microbes visible, huge swaths of other actors played their role, from the microbes themselves, the famous swan neck flasks used in his experiments, the laboratory, the French Academy of Science, and so on.[17] These actors make microbes more real, and, ontologically, science is different with microbes than without microbes. All of which allows Latour to argue that "it is thus possible to say, without contradiction, both 'airborne germs were made up in 1864' and 'they were there all along,' so that 'after 1864 airborne germs were there all along.'"[18] The emergence of microbes thus occurs through an assemblage of support, and after 1864, microbes become an ontologically given reality for the scientists involved in their discovery.

However, no understanding of how performed ontological realities cohere is complete without a corresponding understanding of an assemblage. When Deleuze discusses assemblages, he does so with intense precision, naming territorial, state, capitalist, and nomadic assemblages, though we need not bring forth these structures here. Deleuze and Guattari define "assemblages" as distinct from "strata," and view assemblages as more loosely held, dynamic, precarious, and emergent. For Deleuze, specifically, an assemblage is "a multiplicity which is made up of many heterogeneous terms and which establishes liaisons, relations between them across ages, sexes and reigns—different natures. Thus, the assemblage's only unity is that of co-functioning: it is a symbiosis, a 'sympathy.' It is never filiations which are important but alliances, alloys; these are not successions, lines of descent, but contagions, epidemics, the wind."[19] Despite this complex terminology, we can categorize an assemblage as a set of social and technical relations held in place by shared language conventions, rules, material entities in space, physical relationships, and other entanglements. I tend to think of assemblages as ongoing material and symbolic systems that self-organize, though never as fully stable systems. Helpfully, John Law narrows down Deleuze's possible meanings and turns his description of an assemblage into something a bit more practical, noting that an "assemblage is a process of bundling . . . in which the elements put together are not fixed in shape, do not belong to a larger pre-given list but are constructed at least in part as they are entangled together."[20] Law also notes that Pasteur's microbes can be understood as an ontological phenomenon, wherein they are not confined by discursive representation. Importantly for us, the same can be said for the role of CO_2 in climate change and in geoengineering deliberations—over time, CO_2 has been performed into different realities, from different relations, by different forces, in different assemblages.

Most realities that have been performed into existence overlap to a significant degree. In fact, most things in the world exist in these spaces of overlap, but the few that do not, like CO_2, stand out as currently lacking stasis. The spaces of ontological overlap, of *commensurability*, can be described as persuasive "layers"—not the layers of discursive or sociological analysis laid atop scientific explanations, but the layers of practice-generated realities. These are the layers of different ontologies of practice functioning similarly in different worlds. "Unlike incommensurable theories of languages," S. Scott Graham and Carl Herndl argue, different realities often have places in each other's reality and can therefore become layered.[21] When the overlap of interrelated layers is significant enough, it promotes and encourages alignment between practices. Or to put it another way, *when ontological layers overlap, a functional persuasion is generated.* In medical analysis, Graham and Herndl's context, alignment is clearly the goal, especially in complex diagnoses. Likewise, for science communicators, generating alignment helps convince the public about the seriousness of an issue, as Sean Esbjörn-Hargens has previously explained about climate change.[22]

In the rest of this chapter, I explore how different ontological enactments have generated different versions of CO_2 and then examine potential sites of overlap where further agreement between ontologies could be generated. The categories of CO_2 represented here, and the groups that have produced these versions of CO_2, are not the only ones possible but simply represent some of the most dominant ontological formations. My goal is not to categorize all the ways we disagree about climate change but to show distinctions between several major positions in order to illustrate how arguments for geoengineering are poised to take advantage of these potential overlapping ontological layers. To do this work, we have to go back and evaluate some of the enactments that have generated these overlapping layers; our first stop is the practice of climate modeling and the performance of CO_2 in climate models.

CLIMATE MODEL CO_2

To see how CO_2's ontological status has been defined over time, we need to take a brief tour of what it is, where it comes from, how it is used in climate models, and why it shows up in economic valuations. In chemical terms, CO_2 is a typically odorless, lighter-than-air gas produced when a carbon atom covalently bonds (a bond made from a shared electron pair) with two oxygen atoms. Plants use light to photosynthesize carbohydrates from CO_2 and H_2O, producing O_2 (oxygen); CO_2 is also released whenever any aerobic organism produces energy. It is also produced when organic materials decay and is the most predominant greenhouse gas by volume, after water vapor. The amount of CO_2 in the earth's atmosphere has been increasing in the last few hundred years—from somewhere around 280 ppm to 419 ppm at time of writing.

These measurements fluctuate between seasons, mostly because plant photosynthesis absorbs a substantial amount of CO_2. There is about fifty times as much carbon in the oceans as in the atmosphere, and this oceanic carbon takes the form of carbonic acid, bicarbonate, and carbonate. Increased carbon in the oceans increases the pH, which means oceans have become more acidic in the last few centuries, from approximately 8.25 to about 8.10 (again, at the time of writing). CO_2 is released when carbon-laden materials, or fossil fuels, are extracted from the earth and heated to produce energy. Geologically, fossil fuels are formed when carbon-based organisms live, die, and then decompose over a period of millions of years. Geologic pressure and heat create different states of energy density in these deposits, which is why some fossil fuels give off more energy than others. Likewise, different kinds of life form different kinds of fossil fuels; aquatic phytoplankton form petroleum and natural gas as a result of anerobic decomposition, whereas terrestrial plants tend to form coal and methane.

These details represent the bulk of consensus about CO_2, as the properties and relations that we can identify in other assemblages demonstrate how the compound exists in different ontologies. My goal here is to articulate how climate scientists have produced a CO_2 in climate models and then compare that process to how economists produce a CO_2 that is defined by economic value, in order to identify where instances of persuasive ontological overlap could occur. For understanding how CO_2 is used in climate models, our first stop is 1640, when CO_2 was first "discovered" by Jan Baptist van Helmont, a Flemish chemist credited with developing the term "gas" (from the Greek "chaos") and with determining that a particular gas escaped when he burned charcoal in a closed vessel, though the details of how or why this happened were unclear at the time. Jump ahead to the 1820s, and CO_2 was understood in a more complex way—as key to global temperature and as part of a larger "greenhouse effect." Described in detail by Joseph Fourier, the greenhouse effect was by this time a well-known phenomenon, named for the buildings in which it was first observed. While not an entirely accurate analogy for the atmospheric processes it refers to, the name stuck. In this model, the sun's infrared radiation penetrates the atmosphere and heats the earth's surface, but cannot escape as easily, as it is absorbed by H_2O, CO_2, and other GHGs. The atmosphere reemits some heat into space and retains some to warm the surface of the earth. This general metaphor for how climactic gases function provided the basis for additional experiments and climate modeling.

In older disciplinary formations, climatologists looked for regularities in historical data, in part because the geophysical sciences were not able to perform a controlled experiment—a problem that affects climate science to this day. Not only is there no "control earth," but for a long time, it was simply impossible to take into consideration the huge number of factors that go into a climate model. Svante Arrhenius, frequently named as the originator of CO_2 climate modeling, developed a model that estimated solar radiation and atmospheric absorption to determine what the doubling of CO_2 would do to the earth's global temperature (the earth's "climate sensitivity"). Arrhenius's calculations, carried out in 1896, determined that a doubling of anthropogenic CO_2 emissions would cause the earth's temperature to increase between 5 and 6°C (averaged over all zones of latitude).[23] Following Arrhenius's work, Guy Stewart Callendar built a different climate model, a one-dimensional radiative-connective simulation that divided the atmosphere into twelve vertical levels and accounted for energy transfer between them.[24] In these early models, CO_2 was understood by way of its hypothetical doubling and the corollary impact of that doubling on global temperature. Other climate scientists would add additional properties to this definition.

Around the same time that Arrhenius was determining climate sensitivity with his mathematical models, the American geologist T. C. Chamberlain began to question the cycle of CO_2, speculating how it was stored and released. He wrote a series of mostly hypothetical papers, starting in 1897, suggesting that earlier ice ages followed a variation in the amount of CO_2 present in the atmosphere. While water vapor was the short-term regulator of temperature, CO_2 was the long-term regulator, and changes in CO_2 levels could generate a feedback loop where either heating or cooling would be self-reinforced by other factors in the climate. This belief conflicted with much more widely held claims that CO_2 was self-regulating. One of the self-regulating factors was thought to be water vapor, because an increase in CO_2 would increase the temperature, which would then increase the amount of water vapor in the atmosphere. Water vapor would overlap CO_2 and block out solar radiation, thereby stabilizing global temperatures in a kind of closed system. However, after World War II, scientists began working with more precise physics and were able to identify gaps in the bands of the atmospheric spectrum between CO_2 and water vapor. These new models determined that "even if water vapor in the lower layers of the atmosphere did entirely block any radiation that could have been absorbed by CO_2, that would not keep the gas from making a difference in the rarified and frigid upper layers. Those layers held very little water vapor anyway."[25] By this point, climate scientists had generated a CO_2 with additional properties: the atmospheric heating potential of CO_2 is not canceled out by water vapor, and changes in the amount of CO_2 in the atmosphere likely promote additional changes that reinforce heating or cooling mechanisms. We can now name further values to our climactic

CO_2: its impact on global temperature is related to its concentration in the atmosphere, it can be defined through its hypothetical doubling, its changes are distributed globally, and deviations in its amount could trigger additional feedback mechanisms, because there are no inherently self-balancing forces in the climate system.

Key to these discursive-material relations were some additional properties of CO_2 that were slowly being added high atop Mauna Loa in Hawaii after the International Geophysical Year (IGY). The IGY was a worldwide earth science–focused collaboration effort that started in the summer of 1957 and ended in 1958. It resulted, among other things, in the launch of Sputnik, the identification of the Van Allen belts, the first permanent structure built in Antarctica, the World Data Center system, and a great deal of evidence for the theory of plate tectonics. The extent of IGY funding was such that Charles Keeling was able to obtain infrared spectrophotometers for the Moana Loa Observatory, in hopes of providing more accurate measurements of atmospheric CO_2 than in existing studies.[26] And indeed, measurements taken at the observatory, the world's oldest continuous CO_2 monitoring station, have verified CO_2 levels in the atmosphere. These records show a steady increase of CO_2, and the "Keeling Curve" has since become one of the most famed data global warming visualizations (more on the Keeling Curve in chapter 5). The amount of atmospheric CO_2 measured at the Mauna Loa Observatory is exhibited in a precise way. Essentially, air is pumped into a cylinder that also allows light in through a transparent window. A detector then records infrared radiation as it passes through the air. More CO_2 means more absorption of infrared light by the air in the cylinder and less to be detected by the sensor. The sensor converts a measurement in volts, a potential energy, into a measurement of CO_2 through "extensive and automated (always ongoing) calibration procedures."[27] Typically, CO_2 is defined in parts per million, a quantity that describes its relative density. Estimations concerning the sheer weight of CO_2 show up as a CO_2 defined by $GtCO_2/yr$, the gigatons of CO_2 that have been released over a period of time. This standard of CO_2 can also be used as a unit of measurement by itself, as other GHGs can be represented in units of $GtCO_2eq$, or equivalent gigatons of CO_2. Finally, CO_2 is identified in its energy value, in part to make a comparison between the sun's ability to heat the surface of the earth and the role of additional CO_2. Here, CO_2 can be kW/h per m^2—a measurement of power produced during one hour within the space of an m^2. All these precise measurements are important to the form of CO_2 generated, in part because they assist in producing a CO_2 that can be employed in climate modeling.

Modern climate models, known as global climate models (GCMs), are computerized multidimensional simulations that replicate climate over time and consider a huge array of variables, including "planetary rotation, insolation, gravitation, heat exchange, humidity, orography (surface features such as mountain ranges), sea surface temperature and friction."[28] Needless to say, the work of modern climate modeling is fantastically complex. Modern GCMs function by means of two main elements: a "dynamical core" that simulates large-scale movement using fluid motion equations, and a "model physics" element, which focuses on all other major processes that occur in the atmosphere, including heat and moisture transfer and friction. The "model physics" elements are some of the most difficult to frame and tend to invite the most scientific disagreement, because climate models do not start from observations. Instead, they simulate a climate that is compared to global data that has been gleaned and translated to be of use to climate modeling. Paul Edwards, responsible for a remarkable history of climate modeling, explains that the model physics portions of a GCM are built from both parameters and "tuning."[29]

Parameters are stand-ins for something that climate scientists cannot yet estimate, like the many radiative interactions among the aerosols in the atmosphere and the complete surface of the earth. Both figures would be possible to calculate with line-by-line models but are simply too massive for most computer models, so they are substituted with parameters based on observational and meteorological data. Any given GCM will contain thousands of parameters. "Tuning" is a different process, often called the "necessary evil" of GCMs.[30] "Tuning means adjusting the values of coefficients and even, sometimes, reconstructing equations in order to produce a better overall model," where "better" is understood to mean that results agree "more closely with observations, or that it corresponds more closely to the modeler's expert judgement" concerning the "physical plausibility" of the change.[31] Certainly, we could connect this kind of "tuning" with that described by Thomas Rickert as being "attuned"—an always-ongoing sense of synchronicity and harmony.[32] Aimee Kendall Roundtree has described this tuning as deeply rhetorical, because scientists make their adjustments based on their attempt to create "a bridge between two worlds: speculation and calculation, theory and experimentation" by way of rhetorical strategies that make and report simulation choices, some of which are ad hoc.[33] This level of modeling requires some qualitative uncertainty—a feature that critics of GCMs have amplified. Perhaps the most noteworthy of these critiques, from Oreskes et al., argues that models reproduce a kind of inductive argument. A model, Oreskes et al. claim, may assist scientific knowledge and "may resonate with nature, but it is not a "real" thing," though they appear to be seeking a form of realism that, frankly, may not exist.[34]

Regardless, the properties amassed by CO_2 through modeling are pretty numerous: the heating potential of CO_2 overrides atmospheric water vapor and thus is not part of a self-balancing system, CO_2 can be evaluated by its hypothetical doubling and impact on global temperature, changes in CO_2 are distributed globally, and changes in the ppm of CO_2 activate feedback mechanisms that also raise global temperatures. Virtually all climate scientists agree that these properties exist and so it can be said that this ontological formation has reached *disciplinary* stasis. The climate models created in the last few decades are remarkably precise with respect to each other and have enacted a remarkably functional and consistent version of CO_2. Not in stasis with this performance, however, is an alternative ontology of CO_2 formed by years of economic practice.

ECONOMIC CO_2

Environmental economics, which functions as an academic field, industry practice, and financial methodology, has developed a unique conception of CO_2 generated from both describing and prescribing how CO_2 functions in financial markets. Environmental economics practitioners study the effects of different financial policies around the world, investigate possibilities for acting upon detrimental environmental and public health causes, and suggest economics-based solutions or mitigation strategies for improving environmental conditions. While the definition of CO_2 that has emerged from these practices includes many of its chemical properties, it also incorporates additional relations, because it exists as part of a much larger financial assemblage. Though this economic CO_2 is linked to older concepts in economics such as natural resource scarcity, externalities, equilibrium, and common property, the modern era of environmental economics began in the 1960s.[35] It emerged from the Western environmental movement, a more intense focus on resource scarcity, and active attention to the ecological limits of economic activity. In this era, arguably the most influential work was Kenneth Bouding's "The Economics of the Coming Spaceship Earth", which predated Buckminster Fuller's similarly titled book by two years.[36] Boulding's conception of a spaceship Earth framed the planet as a closed system that can only be restored through solar power and is damaged by overconsumption and abuse. He positioned what he called "cowboy economies" against "spaceman economies," with cowboys assuming limitless resources and spacemen acting as though the planet must be cared for if it is to perpetually sustain life. Boulding argued that in a spaceman economy, "what we are primarily concerned with is stock maintenance, and any technological change which results in the maintenance of a given total stock with a lessened throughput (that is, less production and consumption) is clearly a gain."[37] An economic framework that considers the viability of the earth and its limited resources thus took root in economic practice.

Following from Boulding's assertion that the earth is comprised of finite environmental and economic resources, policies were devised to protect specific environmental assets and contain potential feedback loops of negative environmental destruction. Key to these economic policies were definitions of environmental "externalities." In economics, an "externality" is a concept roughly defined as the cost or benefit that impacts a group of people who did not intend to experience that cost or benefit. In the 1960s pollution became the most easily identifiable environmental externality, as the companies that generated pollution distributed its effects across a population that did not agree to that impact, even though some may have been the end-users of the energy sources that resulted in pollution. Governmental evaluations of how much pollution was acceptable led to the first legal and financial regulations for specific pollutants, clean-air regulations, and economic systems that defined tolerable levels of pollutants by both volume and price. These economic regulations, often called cap-and-trade programs, limited emissions of a particular pollutant for a certain sector of the economy and then issued credits (or permits) to emit that pollutant up to a predefined maximum. In the United States, the 1970 Clean Air Act and the 1977 Clean Air Act amendments imposed regulations on sulfur dioxide (SO_2) emissions, though the first national cap-and-trade system for SO_2 was enacted in 1990 (because of concerns over acid rain).[38] The resultant cap-and-trade system allowed power plants to purchase credits to pollute, but they were also economically incentivized to pollute less over time, as allowance credits could be bought and sold within the industry. This economic practice led to a greater awareness of cap-and-trade economic programs as a control for environmental pollutants. In a very real way, these cap-and-trade systems made the climate available for measurement and definition through financial equations that asked energy industries to apply technological changes to their production processes. This economic regulation also set a standard for the amount of pollution that was environmentally and economically "acceptable," thereby controlling the composition of the atmosphere through both symbolic and material means.

While the cap-and-trade system for SO_2 limited some of the destructive effects of acid rain and was soon applied toward other pollutants, cap-and-trade policies for CO_2 have had mixed success. While scientific consensus emerged about the impact of increased atmospheric GHGs and the 1997 Kyoto Protocol called for state-controlled GHG emission reductions, the cap-and-trade policies that have been drafted and put into place resulted in only modest global CO_2 emission reductions. In the wake of the Kyoto Protocol came the International Emissions Trading system (IET). The IET identified target amounts for different countries over a certain time period, whereby allowed emissions were contained by Assigned Amount Units. In generating this economic regulation on atmospheric CO_2, "a new commodity was created in the form of emission reductions or removals. Since carbon dioxide is the principal greenhouse gas, people speak simply of trading in carbon. Carbon is now tracked and traded like any other commodity," and the systems that arose in response now constitute an economic formation of CO_2.[39]

The assemblages surrounding CO_2 economics have given rise to a different CO_2 than that of climate scientists, though their creations do share some features. Instead of measuring CO_2 by mass alone, economic formations of CO_2 are incorporated into a broader framework wherein CO_2 exists as a commodity valuated by different economic markets and weighed against environmental damage translated into financial terms. The European Union Emissions Trading Scheme trades in European Union Allowances, the Clean Development Mechanism measures in Certified Emissions Reduction (CER), and the Joint Implementation scheme assesses CO_2 in specific in Emissions Reduction Units (ERUs). Each of these carbon credit systems, which have no global authority, estimates the weight of GHG emissions, defined as a measurement of tons of CO_2 (or tCO_2) and correlates weight to "global warming potential" (GWP). GWP has helped link ERU or CER units, in terms of one metric ton of CO_2 equivalent, using the GWP of GHGs as specified in the Kyoto Protocol.[40] The measurement of GWP, however, is more than an epistemic representation of CO_2: it is indicative of an ontological formation. Luigi Pellizzoni argues that GWP (and its measurement) is "a physical entity that comes into existence through complex sociotechnical work."[41] The materiality of CO_2 must be produced and sustained by numerous actors within a world in which it is able to generate new associations. In environmental economics, the mass-based economic definition of CO_2 is "totally absorbed by its commodity status," where it "oscillates not only between calculation and speculation but also between reality and conceptuality, matter and information, the ontological and the epistemic."[42] Here, CO_2 becomes a chemical compound defined by weight but performed as a calculated material construction with numerous other economic properties reacting within existing financial systems.

Regardless of how CO_2 has been formulated in this economic ontology, however, the success of these economic systems—both on a global and local scale—has been unremarkable. Some countries have been able to lower their CO_2 emissions with the European Union's Emissions Trading System, but the overall effect on the climate is disheartening. In 2010 global CO_2 emissions reached 33 $GtCO_2$. In 2018 over 38 $GtCO_2$ were released, a new record.[43] The 2018 *United Nations Emissions Gap Report* explained that unless countries strengthen their 2030 climate commitments, global warming cannot be limited to 1.5°C and that it is extremely unlikely warming can be limited to below 2°C. That same year, the IPCC special report titled *Global Warming of 1.5°C* explained that global warming is *likely* to reach 1.5°C between 2030 and 2052, and that to limit global warming to below 2°C, CO_2 emissions need to reach net zero around 2070.[44] The reach of cap-and-trade systems is not significant enough, nor are the guidelines set by individual cap-and-trade systems strict enough, to reach these goals. Simply put, there are not sufficient economic systems in place to curtail global carbon emissions to avoid additional, potentially catastrophic, global warming. In the United States, which has no national cap-and-trade system but does calculate the social cost of carbon, we can point to the failure of several cap-and-trade bills, most significantly the 2009 American Clean Energy and Security Act, or Waxman-Markey Bill.[45] Whether that bill would have succeeded in helping lower global carbon emissions is perhaps moot, because the properties of CO_2 in an economic ontology are not supported by all economic formulations in the United States; nor are they unanimously believed to be advantageous to continued economic growth. Economists, unlike climate scientists, are not in interdisciplinary stasis about how CO_2 performs. In the West, fiscally conservative think tanks, lobbyists, and legislators often decry the use of carbon markets as creating negative economic impacts.

Indeed, an environmental economic ontology has created a CO_2 that is largely defined by its position in a commercial and industrial assemblage, and unfortunately too many see the cost of restricting CO_2 emissions as an economic disadvantage. Staunch fiscal conservatives define CO_2 in terms of energy that is explicitly subservient to the economic relations in which CO_2 operates. These economists rarely deny the basic attributes of CO_2 that climate scientists have generated; nor do they deny that scientific evidence shows a consistent rise in global temperature that will have implications for both GDP and national security. Instead, they see the economic systems that could be put into place to mitigate global warming, like cap-and-trade, as harmful to the economies in which CO_2 is ultimately an insignificant force. Typically, this group questions the extent of global warming because of the financial impact that cap-and-trade would instigate. Jacques, Dunlap, and Freeman's work on conservative think tanks' responses to environmental regulation identifies a general dismissiveness toward the seriousness of environmental problems, particularly as those problems suggest greater corporate regulation or impact a larger narrative of continual economic progress.[46] Groups like the Copenhagen Consensus Center, the Hoover Institution, the CATO Institute, and the Liberty Fund have released papers that pit the cost of curtailing CO_2 emissions against the economic harm of that action. These groups argue that building cap-and-trade policies, subsidizing renewable energy resources, and investing in alternative transportation are economically harmful. In this set of relations, the price for global warming mitigation measures has not yet been justified.

What these climate modeling and environmental economic performances have been able to generate, then, are two forms of CO_2—not two radically distinct compounds, but two incompatible objects built from differing practices. For climate scientists, CO_2 has been identified through empirical study and asked to perform in climate models, but for environmental economists, and especially fiscally conservative capitalist economists, CO_2 operates within the conditions of financial markets and is restricted by capitalism's need for ever-increasing economic growth. The two formations share some features but also differ significantly enough for us to call them ontologically incompatible (though not epistemically incommensurate). Presumably, if they were compatible, we would have seen greater progress on international cap-and-trade projects. Still, even while these formations of CO_2 are not in stasis, we can point to the potential for surprising overlap that could create a persuasive advantage for SRM geoengineering. To see this persuasion a bit more clearly, we need to trace a few additional relations that have been generated by another group with an even more widely divergent opinion on the reality of global warming—a group that could use the lack of stasis on global warming to create stasis on geoengineering.

DENIALIST CO_2 AND THE PINATUBO OPTION

Significant overlap between the forms of CO_2 that climate scientists and environmental economists have generated is now emerging, one that could easily give rise to a strange sort of opportunity for geoengineering to "solve" climate problems. Some in the scientific community are claiming that geoengineering is a feasible option if no meaningful, long-term mitigation strategies can be deployed, and several mainstream scientific bodies have encouraged additional geoengineering research. The 2014 landmark IPCC report was preceded by a 2011 Expert Meeting on Geoengineering that advocated for clear definitions and for developing a framework of assessment and comparison. The final report from this meeting detailed CDR and SRM techniques and explained that geoengineering could be part of the future of climate change adaptation, stating that "possible emergency responses such as 'geoengineering' in the face of extreme climate impacts" could be employed because their "ability to quickly offset warming could help limit some of the most extreme climate impacts."[47] To borrow a phrase from *Scientific American*, geoengineering "is now firmly on the scientific agenda."[48]

We can see this agenda play out in the numerous research projects that have been created throughout the twenty-first century—including the Reef Restoration and Adaptation Program mentioned earlier. These efforts include US-based geoengineering research funded through universities, including projects at Harvard, Cornell, the Carnegie Institution for Science, Rutgers, the University of Montana, and the University of Washington.[49] Other international projects go by an array of acronyms and abbreviations (EuTRACE, IMPLICC, MIGP, IAGP, DFG, and GeoMIP, among others). These are all multi-institutional, multidisciplinary efforts attempting to better understand the development, technological requirements, costs, ethics, and environmental implications of specific geoengineering proposals. Many of these studies attempt to estimate specific yearly figures for geoengineering plans, for example the much-bandied claim that some methods could offset global warming for a cost of as little as $375 million/yr.[50] Studies have analyzed a range of geoengineering proposals, from DAC to albedo enhancement, and have pointed out the distinction between these hypothetical yearly costs and the long-term economic consequences of climate change. While these comparison projects are useful in identifying the kinds of geoengineering possibilities that exist, they also map geoengineering onto claims about CO_2 and have generated additional persuasive momentum for one geoengineering proposal in particular.

The two versions of CO_2 I have described in this chapter both indicate that CO_2 has changed the climate, though climate scientists are in much stronger agreement that CO_2 is already creating global warming. Economists are in less cohesive agreement and an economic CO_2 has attributes that suggest carbon-mitigating activities could bring more economic harm than the continuance of traditional fossil fuel industries. These two versions of CO_2 could result in a persuasive ontological layering that could generate increased stasis—not stasis about CO_2, but about geoengineering. Graham and Herndl's definition of ontological layering helps re-center this point. Unlike in theories of incommensurability, ontological layering can occur when different objects have a place in each other's performed ontological world. In studying medical practices, Graham and Herndl looked for "practical strategies of alignment, incorporation, coordination, and calibration [that] describe a flexible repertoire of rhetorical processes for layering these practices."[51] Here, I am not concerned with promoting alignment but rather with identifying how that alignment could occur.

For climate scientists, CO_2 causes global warming through its ability to trap solar radiation, where an increased amount of CO_2 leads to greater global temperatures. We conclusively know that the level of atmospheric CO_2 is increasing due to human action but have as of yet done little about it. The environmental "tipping point" toward runaway climate change is either shortly upon us or has already happened. For economists, on the other hand, CO_2 may be increasing and may compound the effects of global warming; however, CO_2 is also defined by its economic value, so an economic "tipping point" has yet to be reached. That point would arrive when severe economic constraints upon CO_2 emissions are warranted. While some economists believe we have already reached it, others do not—and judging by the relatively milquetoast cap-and-trade agreements that currently exist, there is still more concern about protecting GDP than addressing global warming. The opportunity for persuasive overlap will fully emerge when these tipping points align; this point will likely occur when environmental conditions deteriorate to such an extent that many Western economies are economically impacted by climate change and therefore elect to alter CO_2's economic valuation. In this future moment, the two versions of CO_2 will overlap enough to result in a condition where geoengineering can emerge as an agreed-upon means of economic and environmental control. Surprisingly, we do not have to wait to see this overlap happen; for some groups, it has already occurred.

Recently, the expectation of this overlap has caused a strange, incoherent reversal among some climate change deniers, a reversal that has generated additional momentum for SRM geoengineering in particular. Those who completely deny that global warming is happening often fixate upon the sources of doubt in climate models or the ways that the proposed mitigation solutions to excessive GHG output could place Western countries at an economic disadvantage. The specifics of the denialist arguments have been covered by others in much more detail, and it is not my goal here to enumerate the many reasons why a person or group of people could reject scientific consensus.[52] Instead, I want to highlight how climate change denialist organizations have switched from rejecting climate change to simply refuting that climate change is anthropogenic. This shift is evidence of the persuasive ontological layering described earlier in this chapter and is also one of the first indications that geoengineering could gain additional momentum from this future overlap.

Dunlap and McCright have linked the economic-based argument for the denial of climate change to a series of actors who make up what they call the "Climate Change Denial Machine."[53] The latter comprises the fossil fuel industry, corporate America, conservative foundations and think tanks, and reactionary media outlets. In part because they are not involved in generating a CO_2 that can be performed, climate change denialists do not produce CO_2 in the way that climate scientists or economists have. Instead, denialists will often claim that CO_2 has neither the same climatological abilities nor the same scale of impact as climate scientists' CO_2. Their goal is to practice a feigned ignorance in order to sow doubt. Those who claim CO_2 is not increasing—or if it is, that increase does not matter, or if it does matter, it has not been caused by humans—refute the notion that climate models are in any way applicable to "real life." This perspective situates CO_2 with almost none of the same properties of climate scientists' versions. Or to put it another way, a global warming denialist account of CO_2 is based on the *rejection* of attributes. For these groups, either CO_2 is increasing because of processes beyond human control, or it has not increased in atmospheric mass in any substantial way, because CO_2 is only a trace gas in the atmosphere. In the former case, humans cannot exert meaningful change upon the vastness of nature, and in the latter, the ppm measurement of CO_2 functions as evidence of its relative scarcity and inconsequentiality, and therefore its failure to contribute to global warming. Denialist lay logic suggests that anything measured on such a minute scale is unlikely to permanently damage something as massive as the earth. Scientific communities are frequently claimed to be constricted by groupthink, hungry for grant money, in cahoots with liberal politicians, or beholden to environmental special interest groups. This argument frames climate models as misleading math gone awry at best, and intentionally duplicitous schemes to siphon money from other scientific research programs and the taxpayer at worst. Similar opinions have been echoed by conservative groups like the Heritage Institute and the Heartland Institute, among others. They disavow scientific claims about the causal relationship between CO_2 and ecological and economic effects, yet the emerging ontological overlap described previously has resulted in some conservative climate deniers "flipping" perspectives and actively promoting geoengineering projects.

Some of the most strident global warming denying organizations have decided to promote geoengineering, resulting in what some have called the "Super-Freak pivot." This pivot is evident in position papers and policy suggestions and happens when climate change deniers find themselves actively interested in geoengineering strategies that seem antithetical to their functional understanding of CO_2. The name comes from the rather pretentious description of geoengineering in Levitt and Dubner's *SuperFreakonomics.* They describe climate change as having a "relatively small chance of worldwide catastrophe" in their chapter "What Do Al Gore and Mount Pinatubo Have in Common?" but position geoengineering as the reasonable solution to keep in our collective back pockets, just in case scientists turn out to be right.[54] Despite their condescension, their recommendation illustrates where this ontological overlap is generating persuasion.

We can point to several conservative think tanks with strong track records in denying global warming science that have come out in favor of geoengineering research. For example, in 2010 the American Enterprise Institute (AEI) funded a conference on geoengineering named "Evaluating the Geoengineering Option." The event featured speakers who questioned whether geoengineering would provide a beneficial alternative to strategic GHG mitigation policies. According to AEI's materials, conference participants widely agreed that geoengineering is worthy of additional attention and more research is needed.[55] AEI is not alone, either. The aforementioned Copenhagen Consensus Center, the brainchild of climate skeptic Bjørn Lomborg, produced the report *An Analysis of Climate Engineering as a Response to Climate Change*, which asked, "If the global community wants to spend up to, say $250 billion per year over the next 10 years to diminish the adverse effects of climate changes, and to do most good for the world, which solutions would yield the greatest net benefits?—i.e. what are the costs and benefits of different viable climate interventions . . . given some reasonable assumptions about sensible policies for the rest of 21st century?"[56] The general response to this question has been to support geoengineering research. Likewise, the Hoover Institution seems similarly optimistic about geoengineering, even though in 2018 they still claimed that we should "proceed with caution before making bold claims that the main, or even sole, driver of climate change is the human generated increase in the carbon dioxide level" and that "emission controls alone may not stabilize the climate."[57] Their position on whether anything can be done about climate change is blunt: "Yes. We can develop ways to increase the amount of sunlight that is reflected back into space by the upper atmosphere, an approach rarely discussed at global warming conferences. This strategy falls under an approach called climate geoengineering."[58] These positions, and many like them, have come from organizations that have spent years denying that climate change has been occurring, yet are now considering climate change as a kind of economic thought experiment—with geoengineering as a potential solution. The incoherence of this Super-Freak pivot reflects the possibility for the ontological overlap described previously.

But there is a bit more evidence for the emergence of this overlap—this time at the congressional level. In 2017 the US House Science, Space, and Technology Committee's Subcommittee on the Environment discussed geoengineering research ideas. House Republicans attempted to consider the merits of geoengineering research without actually stating that anthropogenic

GHGs are causing climate change. The hearing, according to journalist Brian Kahn, was "at once comical and damn terrifying," as "none of the Republicans could bring themselves to acknowledge that carbon dioxide is the root cause of climate change."[59] Republican congressman Andy Biggs managed to frame the hearing as a discussion on "the viability of geoengineering and any early-stage research associated with this approach. The hearing is not a platform to further the debate about climate change."[60] The mental gymnastics required to hold a hearing on solutions to climate change while actively ignoring the causes of climate change showcases the same logic as the Super-Freak pivot.

Conservative economic conceptions of CO_2 have shifted from a worldview where CO_2 does not impact climate change and where climate change is not occurring to one where global warming could be occurring, irrespective of what properties CO_2 has. This position is no longer one of complete global warming denial, but one where the anthropogenic causes to global warming can still be rejected. As a result, conservative think tanks appear to favor SRM in general and SPI (the "Pinatubo Option") in particular. Much of the conservative think tank interest into SRM techniques comes down to economics, unsurprisingly, and the estimates for implementing SPI are quite low when compared to the potential economic effects of climate change or the costs of GHG mitigation. The Copenhagen Consensus Center's estimates for SPI come to about $22.8 billion/yr., starting in 2025.[61] This number has been roughly confirmed by other studies, as in a US Government Accountability Office report from 2011, and echoed by Levitt and Dubner in the logically challenged *SuperFreakonomics*. While that number might seem astronomic, it pales in comparison to the estimated cost of climate change damage or GHG emission cessation. Plus, SPI includes opportunities for profitable investments in new geoengineering technology. In short, climate change deniers see SPI geoengineering as economically advantageous, because the economy would stand to lose less with last-minute SPI implementation than with other long-term, environmentally sustainable measures, such as a strong carbon taxes or intensive carbon abatement strategies.

Economics aside, there is a second, more troubling reason why SRM techniques are favored among conservative think tanks. For strident climate change deniers, CO_2 does not cause global warming and, regardless, any increases in CO_2 have been nonanthropogenic. SRM geoengineering does not conflict with these beliefs, because an SRM strategy claims to cool the earth *irrespective of whether CO_2 causes global warming or not.* SRM geoengineering does not lower the amount of CO_2 in the atmosphere or the oceans but instead redresses global temperature, which can be ascribed to other, nonanthropogenic causes. It is possible for climate change deniers to hold fast to

Michael Mann's first position—"CO_2 is not actually increasing"—and happily invest in patentable SPI technologies. The incoherence of this position is staggering but unfortunately also viable because of the way that SRM thrives off the potential overlap in formations of CO_2. SRM geoengineering could present itself as a solution that fits how frustrated climate scientists, conservative economists, and reactionary think tanks understand CO_2. Unfortunately, this form of geoengineering also comes burdened with a significant amount of risk.

Plainly put, SPI geoengineering is one of the most alarming climate modification methods currently devised. A 2008 study led by Philip Rasch, chief scientist for climate science at the Pacific Northwest National Laboratory, led to numerous conclusions about the potential risks involved in SPI implementation.[62] The dangers include disruptions to the polar seasons, decreased rainfall patterns, increases in acid rain, surges in ozone depletion, and an intensification in ultraviolet radiation. Likewise, in their review of geoengineering plans, Naomi Vaughan and Timothy Lenton found that "sulphate injection modelling showed disruption to the Asian and African Summer monsoons" and stressed that much more research is required before SPI could be considered a viable geoengineering option.[63] They concluded that "solar radiation management is not an alternative to mitigation, and would have to be deployed in conjunction with carbon dioxide removal geoengineering or maintained on timescales >10,000 years in order to avoid extremely abrupt warming if they failed, or when the intervention stopped."[64] SPI, then, is neither conclusively safe nor conclusively cheap. Substituting aerosol geoengineering for actual GHG abatement strategies is economically ineffective, because of the costs of sustained upkeep and the "deeply uncertain estimates of the damages due to aerosol forcing."[65] With any lapses in continuous particle injection over the next few centuries, environmental risk—and the cost of recovery—becomes too high to be acceptable.

In the emergence of the Super-Freak pivot, then, we have evidence for the beginnings of stasis—not stasis about climate change, but an ontological overlap of performances of CO_2 that help generate stasis about geoengineering. The CO_2 that climate scientists have generated in their modeling efforts may not match the financial one generated by economists, though both can be measured, valued, and tracked in similar ways, and both have an impact upon the systems in which they are a part, whether by amplifying solar radiation or through existing financial systems. Think of this ontological overlap as a kind of Venn diagram, with climate scientists, environmental economists, and others offering discursive-material performances of CO_2, and with SRM geoengineering operating in the center. While conservative think tanks and climate change deniers need not "believe" in either of the more-established versions of CO_2, they could advocate for SPI, pushing it to the forefront of the geoengineering agenda, because it would operate effectively at the moment when environmental and economic tipping align. Further, because of the threat of these potential coordinating tipping points, we are witnessing additional persuasive energy for geoengineering research in the scientific community—they want to generate as much knowledge now, to prepare for when this overlap occurs. As such, we are also witnessing the gradual emergence of geoengineering stasis, and there is additional evidence in the United States for consensus across the ideological divide. In June 2020 Democrats on the House Select Committee for the Climate Crisis released *Solving the Climate Crisis*, a report outlining funding for "decarbonization technologies" and offering suggestions for "climate intervention." While vague, the report suggests that "one possibility is the deployment of atmospheric climate intervention [aka, geoengineering], which aims to reduce global temperature by directly reflecting sunlight away from the Earth, by, for example, injecting aerosols into the stratosphere."[66] Republicans and Democrats are both flirting with SPI geoengineering as a potential solution to climate change, regardless of their position on climate change.

Though unusual, Yuri Izrael's geoengineering experiment aligns all too well with the ontological layers that have been formed over time and are increasingly close to achieving greater overlap. Climate scientists have produced a CO_2 that impacts the earth's absorption of solar radiation, where global warming cannot be canceled out by water vapor and where increases in CO_2 concentration promote reinforced feedback mechanisms. Environmental economists have produced a CO_2 that has similar chemical properties, while also being defined by its economic value and performance in financial markets. Their conceptions of CO_2 do not fully align, and other groups, especially conservative climate change deniers, including conservative think tanks, have found it advantageous to dismiss the anthropogenic origins of climate change but still advocate for geoengineering. They are able to do so because geoengineering can exist in the overlap climate science and environmental economics have performed into existence: the future "tipping point" for when both climactic and economic effects from global warming become unbearable. SRM forms of geoengineering could flourish in this overlap, because SRM geoengineering is not tied to the incompatible forms of CO_2—the causes of climate change—but instead acts upon its effects. Future deliberations will have to respond to the momentum of this persuasive formation, especially if they are to confront the risks involved in geoengineering, something the following chapter picks up in greater detail.

3

GEOENGINEERING RISK

It will not come as a surprise to many that most scientists do not welcome geoengineering, seeing it as a necessary evil—at best—and the reasons why are not particularly difficult to grasp.[1] Active, real-world geoengineering carries with it a number of substantial risks, including environmental and ecosystem damage, disrupted global rainfall, the redistribution of species, potential decreased enthusiasm for ongoing CO_2 mitigation efforts, amplified political discord, and the possibility that whatever technology is used will catastrophically fail. Ocean fertilization could cause toxic algal blooms, cloud brightening could disrupt precipitation patterns, direct air CO_2 capture could escape its sequestered location, and an atmospheric array of reflective discs could result in the sky falling in on our heads. Alan Robock, a distinguished environmental scientist and thoroughgoing geoengineering skeptic, identified twenty reasons why SRM alone may be a bad idea. The many dangers include unknown changes to specific, local climates, continued ocean acidification, higher levels of atmospheric CO_2, less power for solar energy, a host of ethical, political, legal, commercial, and technological issues, and perhaps most broadly, the possibility of "enormous environmental damage."[2] Robock et al. have also compared SPI geoengineering to the "natural" analog of past volcanic eruptions. From those comparisons, we learn that SPI might reduce summer precipitation, weaken the African and Indian monsoon seasons, provide less solar energy for terrestrial and marine vegetation, cause a slight increase in acid rain, deplete the ozone layer over the Antarctic, disrupt ecologies with more diffuse radiation, damage airplanes, degrade astronomical observations, and impact shortwave and longwave radiation.[3] On the upside, however, other researchers have suggested it could create pretty sunsets.[4]

If it not already clear, my concern in this chapter is about risk. The risks of SRM are significant, and the fact that these technologies are actively being investigated—and as the last chapter has shown, actively encouraged—should warrant our concern. Some believe these risks could be mitigated through more intense research or that the risks could be worthwhile if their implementation prevents the worst ecological and economic projections of global warming. As a concept, risk has been expressed by evaluations of material harm and/or the social conditions that produce representations of risk for different groups. Geoengineering risk in particular has too often been understood by one-dimensional material definitions and articulated as a kind of output from a cost-benefit analysis of possible responses to climate change. Articulating the risk inherent in geoengineering from the perspective of a geologic rhetoric prompts us to consider the origins of climate change and follow how these effects persist into the present day, understanding that pressure as a form of geosocial persuasion. Ulrich Beck's theory of a risk society provides one way to identify how this risk is both geologic *and* persuasive. According to Beck, our current society is one where risk is diffuse and generalized, impacting everyone. Hundreds of years of industrial modernity and the consequent release of GHGs present themselves to us as risk but also create a pressure to respond. Geoengineering is just one form of response to this geologic history, though it has been advanced largely by what Beck calls "sub-political actors." These actors, including the "geoclique" of entrepreneurial climate scientists—those conducting real-world SPI tests, carrying out SPI climate modeling projects, and releasing influential reports and conference statements—are actively defining geoengineering in our current deliberative moment. These same subpolitical geoclique members attempt to align SPI to "natural" analogs, reference ecomodernist principles to advocate for private investment, and define risk through modeling and simulation efforts. All these activities work with the potential for overlapping ontological layering described in the previous chapter to "lock in" SPI as the de facto geoengineering option. Recognizing how we got to this point, however, first requires us to examine how risk can be described in persuasive, geologic terms.

UNDERSTANDING GEOENGINEERING RISK

Risk can be a notoriously slippery concept to define, though Paul Slovic has been offering a good description for some time. Slovic has led the move to articulate risk as more than a static, rational concept and has described risk by elaborating on the many ways that the idea can be articulated. In a 2002 working paper for the Center for Decision Sciences, Slovic and Elke Weber explained the most common uses of the term:

Risk as a hazard: "Which risks should we rank?"

Risk as probability: "What is the risk of getting AIDS from an infected needle?"

Risk as consequence: "What is the risk of letting your parking meter expire?"

Risk as potential adversity or threat: "How great is the risk of riding a motorcycle?"[5]

Each of these uses emphasizes something unique. A focus on hazard names an event as a danger; a focus on probability concerns likelihood; a focus on consequence weighs outcomes; a focus on threat emphasizes the potential for harm. Most risk scholars use some or all of these concepts and have become comfortable with a broadened definition, moving away from positivist, supposedly objective models, though risk is still often defined as an objective hazard to specific goals in industrial settings. For example, ISO 31000, the international standards guidelines for risk management, defines risk as the "effect of uncertainty on objectives."[6] Some explain risk through an instrumental analysis of the conditions of risk, as defined solely by experts. Here, risk "is often represented using the following formula: Risk = [Probability of an event occurring] x [Consequences of that event]," where "an assessment of risk is the product of an evaluative process at the individual or organisation level."[7] For others, the sticking point is often the frustrating epistemology from which a meaning is proffered. That is, subject matter experts in different fields often define risk from the perspective of management or industry and have difficulty understanding how a public makes sense of and defines risk from a different viewpoint. On one side lies "expert" risk assessment, which encompasses statistical probabilities of risk, economic analysis of harm, comparative risk analysis, engineering data, and scientific assessment models. On the other side exists public, "nonexpert" articulations of risk, which can involve intuitive thinking, embodied beliefs, personal perspectives, cultural dimensions of risk, cultural memories, and considerations of source credibility. When expert assessment distinguishes these two areas, it "simply means that some group or individual has brought specialized knowledge to bear on a risk issue, usually by referring to published scientific literature and by using technical terminology."[8] Both systems of risk definition are important, of course, though partitioning them into two different camps often serves to rationalize claims for the objectivity of expert assessment on the one hand, and the supposed irrationality of risk perception, on the other.

Comparative analyses of geoengineering proposals have likewise focused on risk that is defined and bounded by scientific specialization. For example,

Yuri Izrael's risk analysis of his sulfate aerosol tests examined geoengineering proposals' natural analogs, any "impacts within natural variations," the results of a "possible quick cease" of the technology, and the technology's ability to compensate for CO_2 doubling by 2050.[9] Izrael concluded that "the method of creating in the stratosphere a layer of sulfate aerosol particles of a submicron size and a total mass of several megatons mostly meets all criteria."[10] That kind of instrumental, and optimistic, conclusion about geoengineering risk should be cause for concern, but Izrael's evaluation positions effectiveness against environmental impact, which is more than some studies attempt—a fair number of other risk assessments of geoengineering only include cost-benefit assessments and assumptions about instrumental rationality borrowed from behavioral economics. Cost-benefit analyses are inadequate measures of risk and are clearly bounded by a frame of economic rationality. The most responsible analyses of geoengineering proposals, like that from Ferraro et al., acknowledge that cost-benefit methods are inadequate, not only because of the huge number of unknowns but also because it remains difficult to convert "physical variables as simulated by a climate model into meaningful monetary costs."[11]

Unfortunately, literal cost-benefit analyses that weigh expense against effectiveness are common among geoengineering evaluations. Huttunen and Hildén explain how in most geoengineering assessments, "geoengineering becomes mainly a technical and engineering matter. Potential social, ethical, and political issues are neglected or dealt with in passing."[12] Philip Boyd's commentary on geoengineering for *Nature Geoscience* follows this pattern pretty closely.[13] One of many scholars who define risk through subject matter experts alone, Boyd studies how many early geoengineering cost calculations now require reassessment, as they were likely too optimistic in assuming effectiveness in countering the impacts from excessive CO_2. To factor in risk, Boyd evaluates "anticipated side effects such as acid rain or ozone depletion," "unanticipated side-effects" that are not yet understood, "entirely unexpected side-effects of an unknown nature," and finally whether a plan's effectiveness is able to be verified at all.[14] Attempts to factor in "unknown unknowns" are commendable, though integrating such poorly understood variables into a more instrumental frame seems problematic, at best. This style of ranking geoengineering plans glosses over the sheer volume of unknown effects in what are still largely hypothetical climatological strategies. In their review of geoengineering proposals, Bellamy et al. found that while "expert reviews and opinions dominate the expert-analytic category of geoengineering appraisals," they also "hide a range of subjectivities" that are "often under-explained and unaccounted for. Similarly, the reasoning and methods behind arbitrary rankings for different geoengineering proposal feasibilities lack transparency."[15] The most "objective" analyses of risk, and the most "rational" forms of cost-benefit analysis, are still decidedly subjective, though reliant upon an epistemology that hides the sources of those variables.

Cost-benefit and market-based evaluations of geoengineering proposals advantage those most aligned with current economic conditions; they consider if a proposal can take advantage of patent law, for example, in their pursuit of a profitable business model. Tina Sikka has neatly explained that such market-based analyses tend to deepen neoliberalist assumptions about corporate profits as the markers of efficiency and effectiveness, at the cost of other alternatives.[16] As we will see later in this chapter, assessments of geoengineering that appeal to an economic status quo are likely to gain additional persuasive momentum, insofar as they can find investors, remain optimistic that capitalism will drive technical and ecological innovation, and define risk by narrow constraints. Rhetoricians have critiqued these models of risk in the past and their analyses show such evaluations as flawed and incomplete, though these same rhetorical claims are mostly aligned with an epistemic, cultural model of risk analysis and communication. This epistemological frame can be ill-suited to studying geoengineering, because it often assumes a Holocene fallacy.

RHETORICS OF RISK

Rhetoricians studying risk have critiqued instrumental or expert-dependent formations of risk in the past, often by bringing risk analysis and communication models into conversation with existing rhetorical concepts. Rhetorical studies of risk focus on how "public" articulations of risk are formed through nonrational rhetorics, personal perspectives, cultural dimensions of risk, and assessments of credibility. For example, Katz and Miller's investigation of a waste site in North Carolina criticized risk communication as too focused on correcting the public's perceptions of risk "so that they would better match the "risk analyses" made by the experts," where the public's perception of risk was understood to be "subjective, mistaken, emotional, and even irrational."[17] In this case study, authority and correct knowledge were almost wholly controlled by the technical experts in charge of the decision-making process. Katz and Miller called this an "engineering model" of communication that can only account for rational subjects. In rational choice models of risk communication, subject-matter experts assume control over the knowledge of risk and their attempts at public dialogue often become opportunities to educate and inform the public about what is assumed to be best for them; effective risk communication is believed to result in public agreement with preformed risk analysis models. This one-way structure misses much of how the public actively participates in the formation of what counts as risk.

The psychometric paradigm of risk offers an alternate theory of risk based on cognitive understandings of hazards, with the broader assumption that risk could be identified through empirical research and applied to different risk situations. Slovic's work addresses both "objective" risk analysis and the cultural dimensions of risk that are generated by people, through discourse, in everyday life. This psychometric paradigm identifies broad ideologically and culturally oriented heuristics that were refined by other researchers and applied to a range of risk contexts. These nonspecific and value-based articulations of "cultural" factors have also been questioned by rhetoric scholars, but with less fervor. Grabill and Simmons's work called out risk communication as being investigated by arhetorical and decontextualized models and argued that these conceptions of risk still gave subject matter experts control. In their view, these methods employed heuristics that help exerts "determine how best to adapt their initial message to the public and how to negotiate the decision making process" at the expense of focusing "primarily on the risk itself, not the public or on a range of other contextual issues."[18] Grabill and Simmons's goal was to situate risk communication within a growing epistemic approach, a rhetoric of inquiry paradigm similar to that which fostered the rhetoric of science. By adapting to these epistemic, social constructionist assumptions, risk communication could gain a much-needed focus on power, local dynamics, emotion, and credibility. Grabill and Simmons reasoned that this critical formation of risk communication could counter "the tendency to develop "scientific" and universal models of assessment and communication that treat risk algorithmically and audiences as universal, rational, and therefore silent."[19]

Most rhetorical work on risk is aligned with an epistemic, cultural model of risk, at least insofar as local, qualitative work has been used to help clarify rhetorical features of the approaches to risk that have already been identified. Cultural theories of risk are much more anthropological in scope and tone, with an understanding of the variables that can impact a group's understanding of what risk means in different situations. Mary Douglas and Aaron Wildavsky, particularly in their 1982 book *Risk and Culture*, defined this approach's attention to cultural and institutional rather than psychological or individual conceptions of risk.[20] Douglas's work discussed "how institutions develop mechanisms of accountability and blame allocation when 'misfortunes' happen" and Wildavsky's research asked, "what leads different individuals to perceive risks differently?"[21] From this model emerged a more "locally based and discursively grounded empirical approach to studying environmental risk."[22] Unlike the engineering or psychometric paradigms, empirical research in this model is often more qualitative, in situ, and less concerned with identifying large-scale patterns. Most rhetorical work on risk aligns with this cultural model, as local, qualitative research has helped nuance the multiple ways that groups define risk.

Research in the rhetoric of risk has contributed a great deal to the understanding of risk definition and communication, yet there remains an ill-fit between these rhetorical methods of risk analysis and the kinds of risks inherent in geoengineering efforts. Sauer's work on mining risk, Simmons's work on environmental risk, Frost's research on transcultural risk, and Potts's research on social media risk, among other examples, have reinforced how rhetorical efforts can be used to advocate for the safety of those most affected by risk.[23] However, one reason that cultural construction models can be a poor fit for understanding geoengineering is because the public is largely unacquainted with the concept of geoengineering. Qualitative studies have sought to determine which geoengineering plans could garner public support, but researchers have found that substantial priming about geoengineering is necessary before survey participants understands the questions. Corner and Pidgeon summarized the public's knowledge of geoengineering in 2015 as "very low," in a review of existing public surveys.[24] Along the same lines, Scheer and Renn noted that when most respondents explained what geoengineering is, they confused it with weather modification or geothermal energy.[25] Reasons for this relatively low level of understanding are easy enough to determine, as very few geoengineering technologies have "stuck" in national news cycles or have been made publicly visible.

Because geoengineering technologies and their ongoing research activities are not something that can be seen, touched, or watched in action by most of the public, they are often still seen as sociotechnical "imaginaries," in Jasanoff and Kim's terminology. Sociotechnical imaginaries are "collectively imagined forms of social life and social order reflected in the design and fulfillment of nation-specific scientific and/or technological projects."[26] When they are understood, geoengineering technologies are somewhat mythical future states or realities that have yet to come to pass—they are as much defined by people's visions, symbols, and associated feelings about climate change or the environment as they are about the technologies themselves. Their abstraction means that for most members of the public geoengineering exists in a state of unreality, and it means that the risks involved in geoengineering are often hypothetical and situated below other forms of risk. In contrast with risk awareness about earthquakes, for example, we can assume that most people who have strong opinions about earthquake risks live in a relatively earthquake-prone area. Those who live in San Francisco, for example, would have what is called a "high familiarity/low dread" model for that risk.[27] That is, they are generally familiar with the kind of event, through their own experience, the experiences of others, or media depictions. Familiarity is often high, and fear is relatively low, because those events occur infrequently, their consequences are generally unavoidable, and the damage they can cause is understood fairly well. In contrast, a low familiarity/low dread model is usually reserved for chemical, biological, radiological, or nuclear accidents.[28] When a population understands a risk through this framework, their knowledge of the event's potential significance is lower, their impulse to prepare less substantial, and their recognition of event likelihood more abstract. We could place geoengineering in a similar category, perhaps a "low familiarity/variable dread model," with "variable dread" signifying the variety of understandings the public has toward climate change, the reasons why geoengineering could be implemented, and the effects of its implementation.

A second reason why cultural construction theories of risk are ill-suited to understanding geoengineering risk is that they often replicate a Holocene fallacy. When risk is defined through as the discursive construction of public perception, it can exist wholly within the horizontal, critical zone of human habitation. Likewise, geology does not often intrude in typical rhetorical constructions of risk, unless it arrives in the form of a natural disaster, and then only insofar as an immediate material consequence or obstacle. In these models, the long history of human interaction with planetary forces may not be addressed, and geology could easily be seen as a surface effect, not a history of discursive-material production or a form of geologic *and* cultural risk. Understanding risk in geoengineering requires a different rhetoric of risk and a different understanding of how material persuasion functions alongside social strata. While not perfect, Ulrich Beck's theories of risk showcase how geology plays a substantive role in today's geoengineering deliberations.

GEOLOGIC RHETORIC IN A RISK SOCIETY

Ulrich Beck's ideas, most notably those in *Risk Society: Towards a New Modernity*, detail how risk has become a pervasive force in Western culture and help explain the current entrepreneurial efforts toward climate change risk-abatement. [29] Beck argued that a new form of society emerged in late modernity—that of reflexive modernity—wherein risk takes on distinct cultural importance. A reflexive modernity is a description of a society that has achieved democracy and created widespread political rights, yet is still impacted by the sociocultural traditions, social norms, and material harms associated with the previous form of modernity, those established in a largely undemocratic, presuffrage, slave-owning, worker-exploiting, industrialization. "Reflexivity," according to Beck, is thus the "autonomous, undesired and unseen, transition from industrial to risk society that cannot be dealt with and assimilated in the system of industrial society."[30] Beck was not particularly concerned with describing modernity as a state of being never actually reached, à la Latour, and more interested in describing how a reflexive modernity is comprised of *reflexive actions* to previous sociocultural formations, as a result of modernity. In a reflexive modernity, "components of a traditionality inherent in industrialism are inscribed in varied ways within the architecture of industrial society—in the patterns of 'classes', 'nuclear family', 'professional work', or in the understanding of 'science,' 'progress,' 'democracy'—and their foundations begin to crumble and disintegrate in the reflexivity of modernization. Strange as it might sound, the epochal irritations aroused by this are all results not of the crisis but of the success of modernization. It is successful even against its own industrial assumptions and limitations."[31] This new modernization aligns with Western, neoliberal ideals that stand in direct contrast to earlier, industrial forms of modernity, undercutting the core "formations of class, stratum, occupation, sex roles, nuclear family," capitalism, and industry that defined and structured modernity.[32] For Beck, reflexive modernity is a stage where progress (illusory or not) can quickly turn self-destructive.

While industrial modernity manifests in a belief in progress, science, democracy, and Enlightenment ideals, in a reflexive society those sociocultural traditions are reflexively re-embedded in ways that lead them to break down over time. Because society is impacted by both the outgoing societal norms of industrialized modernity *and* the risks inherited from that industrialization, Beck argues that Western culture is now in a transitional period where managing risk and being hyperaware of risk are the de facto states of being. The problems concerning the distribution of goods like food, employment, and healthcare have been resolved for some (though certainly not all—Beck was far too optimistic), and the West is now left with the question of how the risks of modernity could be "prevented, minimized, dramatized, or channeled."[33] Risk awareness, negotiation, and avoidance becomes endemic to life itself, where no person can make the choice to avoid risk but instead makes culturally specific choices in negotiation with the risks most central to their values and existence. According to Beck, the central problem of Western society is not in securing the necessities of life, as it was in industrial modernity, but the management of risk, where all people are "caught up in defensive battles of various types, anticipating the hostile substances in one's manner of living."[34]

Few rhetoricians have engaged Beck's theory about risk, perhaps because his research focuses on risk through a broad societal description, where there is little room for the nuance of interpersonal or discursive relations between people or organizations. Beck articulates risk as both ontologically real and socially constructed, whereas rhetoricians have approached risk through a framework of social construction. J. Blake Scott has found the same to be true, suggesting that "rhetorical studies have been chiefly concerned with rethinking risk as a socially constructed rather than objective phenomenon and with rethinking risk assessment and communication as messy rhetorical enterprises rather than more straightforward, scientific ones."[35] Robert Danisch, one of the few rhetoricians to take seriously Beck's central thesis, argues that "for the most part, rhetorical studies has been concerned with elaborating the ways in which risk is socially constructed and demonstrating how risk communication is not a straightforward, technocratic project but a difficult rhetorical enterprise that requires many different actors."[36] Beck's risk society thesis offers a great deal to a geologic rhetoric concerned with deliberation, because reflexive modernity is now impacted by industrial modernity's previous abuses of the earth, which has both material and discursive dimensions. Having been mined for fossil fuels, the earth now distributes the effects of that manipulation throughout the world as increased atmospheric CO_2. Environmental risk is therefore a consequence of industrial resource extraction and constitutes "irreversible threats to the life of plants, animals, and human beings" regardless of economic status or political power.[37] Danisch explains that for a risk society, the chief dangers are environmental problems like global warming—threats that are real and unprecedented in terms of scale and scope, because it is the "magnitude and global nature of contemporary risks that set them apart from earlier forms of risk and lead to the development of a different perception of risk."[38] Deborah Lupton agrees, noting how "contemporary hazards now are often open-ended events, rather than events that have a foreseeable end," leading to apocalyptic-level hazards that threaten life on earth.[39] Global warming influences every person on earth (albeit unequally) and is both diffuse and pernicious.

A large part of how reflexive modernity is *reflexively persuasive* involves our current Anthropocene responses to inherited forms of energy production and relations with geology. While there are numerous possibilities for the origins of the Anthropocene (many ideologically freighted), one suggested starting point is the emergence of the Industrial Revolution and the development of the steam engine in 1784.[40] Dipesh Chakrabarty notes that this period includes when humans moved from renewable fuels to fossil fuels, including coal, oil, and then gas. Western culture has been built upon fossil fuel exploitation, where "the mansion of modern freedoms stands on an ever-expanding base of fossil-fuel use."[41] Our world is dominated and increasingly defined by the long history of this dependence and the resultant release of CO_2; we have inherited a fossil fuel industry that intrinsically supports many deeply structured, ongoing social relations. For ethicist Stephen Gardiner, climate change is "seriously backloaded," insofar as current global warming "is primarily the result of emissions from some time in the past, rather than current emissions."[42] Current life in the West is so thoroughly dependent upon fossil fuel geologic relations that interactions with their longstanding and lingering forces comprise *the quintessential articulation of reflexive modernity*. The extraction of fossil fuels powers the institutions of modernity to which we reflexively define ourselves. Western culture's past and current geologic abuses, and therefore their risks, are now the most substantial persuasive forces upon life on earth, especially as they continue to weigh on any future environmental deliberations we may make.

While the dangers of the industrial era were known because they "assaulted the nose or the eyes and were thus perceptible to the senses," industrial modernity literally and figuratively opened the earth to risks that "escape perception and are localized in the sphere of physical and chemical formulas."[43] As Andreas Malm explains in *Fossil Capital*, our current-day climate emergency is a result of past actions, "the transfer of carbon from geological reserves to fireplaces and thence to the atmosphere, into the running carbon cycle from which it was locked away for ages and eras, sets the process in motion. But the effects are always delayed. It takes time before a certain quantity of CO_2 emissions is realised as a corresponding amount of warming, and before that warming takes its full toll on the ecosystems. For every emission added to past output, the atmospheric concentration of the gas increases, its effect further augmented in accordance with 'the fundamental tenet of climate science: emissions are cumulative.'"[44] Industrial modernity and its concomitant dependence upon fossil fuels both formed the basis for reflexive modernity and generated the conditions for the largest environmental risk to which reflexive modernity is now forced to respond. Beck considers these human-generated risks "manufactured uncertainties"—risks dependent upon human decisions, created by society itself, "immanent to society and thus externalizable, collectively imposed and thus individually unavoidable."[45] Because past geologic extractions established an energy regime that still powers the present, we can identify how this persuasive, geologic, material formation is also a cultural configuration of risk.

As the forces of these (past and present) geologic abuses continue to result in increasing global temperatures, we witness the intermingling of the geologic and the social—exactly the forms of persuasion that Deleuze and

Guattari described. Years of Western fossil fuel extraction have resulted in diffuse and malignant modern-day ecological risk, the broadly defined "risk society" of Beck's analysis. However, Deleuze and Guattari describe the geologic as a layer of intensities and energies formed through sedimentation, where sedimentation is the interaction between the geologic strata and the social strata, a series of relations that Clark and Yusoff term "geosocial."[46] Geosocial formations are found where the geologic interacts with the social and where geologic strata offer up an ungrounded ground upon which social life exists. Echoing out over several centuries of continuous CO_2 output, the geologic effects of industrial modernity contribute enormously to the emergence of climate change as a diffuse, uncontainable, and persuasive risk, the social responses to which must necessarily intermingle with the material ones humans have extracted from the ground. This is a different conception of risk than that proffered by a subject matter expert definition, a psychometric paradigm, a cultural model, or a rhetorical/cultural construction. This is a geosocial formation of persuasive risk, in which Deleuze and Guattari's strata align with Beck's risk society thesis. Geosocial risk is diffuse, uncontainable, and caused by the continued release of energy from the earth that now influences the conditions for life. Additionally, geosocial risk is persuasive because the geologic exerts pressure on the social in influential and persistent ways.

Geosocial risk persuades in ways that might not immediately recognizable, yet it can be disguised as simply as the material effects of pervious geologic abuse impacting current generations. Certainly, the CO_2 in our present atmosphere and the climactic changes we are currently experiencing attest to this risk. Yet, the material effects are, as Deleuze and Guattari suggest, always distributed in different strata, because the geologic strata are made up of social forces, and social strata are influenced by geologic pressures. If we leave the geologic and the social strata as distinct in our conception of risk, we fail to understand how the geologic is inherent in, and forms the always-unstable basis for, all social relations. We make the mistake, once again, of ignoring verticality. Kathryn Yusoff explains that in order to understand the geosocial, we need to remember that for Deleuze and Guattari "stratification is the structural axis for the organization of social and geologic formations that can be extended into an analysis of geosocial formations."[47] What we witness in the emergence of geosocial risk, then, is the organization of forces from each stratum pushing through into others, where (in Deleuze and Guattari's terms) the inorganic (geologic) stratum is part of the organic (biological) stratum and the alloplastic (social) stratum. This geosocial articulation of risk recognizes how the social stratum and the geologic stratum have changed each other and resulted in the disruptions we currently experience. This geosocial formation of risk also identifies how capitalism, specifically, both functions as the prime generator of geosocial risk and is currently acting as the leading force in responding to this risk.

Certainly, climate change understood as a geosocial risk puts the onus of responsibility on capitalism. We can think of capitalist economics, and its attendant forces of neoliberal state governance, colonialism, and resource extraction, as what Yusoff calls a "geosocial machine," one good at organizing "modes of capture that capitalize on geopower."[48] Specifically, capitalism-as-a-geosocial machine has affected changes in the geologic stratum by forcing extractions of all varieties. The result of those extractions—of coal, oil, gas, minerals, and other elements—creates enormous geosocial risk for modern-day humans and nonhuman animals in the form of climate change. The agents of capitalism, then, have caused this significant amplification of geosocial forces, resulting in risks that now threaten life, but will likely continue until stability can again be reached. Understanding this machine will also help us explain why, to confront these pressures, institutions outside of traditional governmental bodies have begun to emerge—groups that Beck calls "subpolitical actors."

SUBPOLITICAL ACTORS

As the geologic forces of industrial modernity wreak havoc on our ecosystems and lives, they also put persuasive pressure on us to respond via the geosocial rhetorical forces at play in reflexive modernity. At the same time, however, we are witnessing a weakening ability of state-level agencies to make deliberative decisions about these circumstances. We do not need to look far for examples of these risks, as aggressive forest fires, worsening droughts, increasing sea levels, and amplified storm flooding are just a few of the ways that geologic forces impact human and nonhuman life. Safety from events like these is increasingly unlikely to be assured by governments, local or federal, a condition Tina Sikka describes as one where the "the logical classification of risks are . . . rendered impossible."[49] The formal institutions of reflexive modernity are unable to cope with the powerful geologic consequences of industrial modernity, though they may try, by scattershot methods, to do so. Increasingly, we see an increase in subpolitical actors active in these endeavors—institutions, companies, or agencies that exist below the level of independent nation-states or state-funded organizations. The result is the emergence of groups that operate independently of official governmental (and sometimes legal) restraints. Beck has argued that "only a part of the decision-making competencies that structure society are gathered together in the political system and subjected to the principles of parliamentary democracy. Another part is removed from the rules of public inspection and justification and delegated to the freedom of investment of enterprises and the freedom of research of science. Social changes in these contexts are displaced as latent side effects of scientific and technoscientific decisions. People do something quite different: they assert themselves in the market, use the rules of profit-making, carry forth scientific and technical inquiry, and in so doing they turn over the conditions of everyday life."[50] These subpolitical actors represent a "system of scientific, technological, and economic modernization" and have assumed a leading role in advancing technology, shaping society, and responding to risk from explicitly political actors.[51] In a risk society, risk itself is multifaceted and broadly manifest, so a range of innovative research and commercially oriented subpolitical actors can attempt to provide resources that ameliorate risk, and thrive by doing so. Geoengineering has so far been dominated by these actors—capitalist enterprises willing to capitalize on the geologic risks of reflexive modernity.

When traditional institutions of power are unable to cope with the geologic reflexivity, a power vacuum emerges and entrepreneurial subpolitical actors from all potential political viewpoints are able to rush in and fill this void. The logic of these subpolitical projects resembles what Sikka calls "the kind of technophilic embrace of geoengineering like Richard Branson's 'Carbon War Room' which he recently formed with individuals from government, universities and business and policy circles with the aim of implementing market-based solutions to climate change. . . . Branson maintains that it is in the realm of technological innovation that solutions to the climate problem will be found."[52] Critical theorist Imre Szeman agrees, writing that "we all live in a world in which the unquestioned social value and legitimacy of entrepreneurship shapes public policy, social development, economic futures, and cultural beliefs and expectations" and thus entrepreneurial subpolitical forces are able to gain credibility and power.[53] We can identify these subpolitical actors in the research collaboratives that are studying geoengineering technologies, the Silicon Valley–style start-ups that are testing potential solutions, and even the nongovernmental organizations that are creating geoengineering policy statements. Companies like Intellectual Ventures have developed widespread geoengineering interests, ones like Global Thermostat and Carbon Engineering have focused on CO_2 capture and storage, and numerous others are trying to produce marketable technologies by way of cloud brightening, biochar, or ocean fertilization.

One of the most influential subpolitical actors for geoengineering has not been a person at all, but the 2010 Asilomar International Conference on Climate Change Intervention Technologies, as this conference jump-started numerous commercial climate engineering efforts. The event brought together more than 175 scientists, engineers, lawyers, environmentalists, and policy experts to consider the risks associated with geoengineering. At its conclusion, attendees agreed upon a public statement, one that reaffirmed their commitment to the development of low-carbon energy sources and acknowledged the insufficiency of human efforts to reduce greenhouse gas emissions. The statement emphasized that it is "important to initiate further research in the natural and social sciences to better understand and communicate whether alternative strategies to moderate future climate change are, or are not, viable, appropriate and ethical."[54] The statement's authors proffered a unified voice arguing that further research into geoengineering was necessary, because "we do not yet have sufficient knowledge of the risks associated with using methods for climate intervention and remediation, their intended and unintended impacts, and their efficacy in reducing the rate of climatic change to assess whether they should or should not be implemented."[55] Although it is just one example of the substantial networks being built around the concept of SPI, this conference was the first influential gathering of geoengineering researchers known as the geoclique.

Notably, the Asilomar conference was attended by Paul Crutzen, a key member of the geoclique, a term deployed by science writer Eli Kintisch to denote a group of subpolitical actors comprising scientists and entrepreneurs in favor of increasing both publicly and privately funded geoengineering research projects. The geoclique controls much of the funds for geoengineering "via a private trust," bringing together "young researchers from around the world to swap ideas under their auspices, and their own patents and investments in geoengineering technologies."[56] Clive Hamilton argues that these researchers, investors, and policy experts have managed to define not only the general scope for geoengineering research but the kinds of risks associated with SPI. The geoclique has "assumed a privileged place in advising not merely on technical questions but on governance arrangements, ethical concerns and international negotiations, despite their lack of expertise. There is a view that if you are clever enough to understand atmospheric physics then you are clever enough to grasp the nuances of politics, social change and ethics."[57] The clique has encouraged private, commercial investment into geoengineering and has attempted to frame public opinion, for example, in discouraging the use of the terms "solar radiation management" and "geoengineering" in favor of "sunlight reflection methods" and "climate remediation."[58] Crutzen, along with David Keith and Ken Caldeira, have been the loudest voices in the geoclique, as both Keith and Caldeira participated in the landmark 2009 Royal Society report, contributed to a UK House of Commons inquiry, and testified to the US Congress about the need for government-funded geoengineering research. Hamilton believes their views have also helped "frame the deliberations of the Intergovernmental Panel on Climate Change as it grapples for the first time with the scientific and ethical tangle that is climate engineering."[59] As I suggested earlier, climate science, environmental economics, and climate denialist perspectives on CO_2 indicate that a persuasive layering is coming, one that could result in a state of exception wherein SRM geoengineering is the leading option. Here, the actions of the geoclique have followed suit, pushing SPI to the front of the geoengineering agenda and prompting further research to understand its potential impact on our climate future. Representing a subpolitical force driving inquiry, investment, and discussion on geoengineering, the geoclique has also been able to operate just outside of official international environmental agreements.

Indeed, subpolitical actors have proven difficult to contain through existing environmental treaties, as the only applicable ones were drawn up well before geoengineering emerged. The US Government Accountability Office, for example, has struggled to match up existing agreements to SPI, suggesting that the 1967 Treaty on Principles Governing the Activities of States in the Exploration and Use of Outer Space could apply, as could the 1985 Vienna Convention for the Protection of the Ozone Layer.[60] Likewise, the influential Royal Society has admitted there is no international agreement that effectively governs the use of the atmosphere. The society's 2009 report notes that the "injection of aerosols is subject to the jurisdiction and control of the sovereign whose air space it is injected into," though a 1979 agreement, the Long-Range Transboundary Air Pollution Convention, might be applicable.[61] Without a legal test, though, these are just educated guesses. Other possibilities include the applicability of other international agreements, ranging from the 1976 Convention on the Prohibition of Military or Any Other Hostile Use of Environmental Modification Techniques, the 1987 Montreal Protocol, and the 1992 UN Framework Convention on Climate Change.[62] Frankly, these speculations will stay hypothetical for the time being, and the official legality of any geoengineering trial or test, including Yuri Izrael's aforementioned experiment, is remains remarkably murky.

The IPCC 2014 Climate Change Synthesis Report frames the official governance context for geoengineering rather mildly, suggesting that SRM strategies raise "questions about risks, governance and ethical implications of development and deployment" and that there are "special challenges emerging for international institutions and mechanisms that could coordinate research and possibly restrain testing and deployment."[63] These methods are imprecise and would require, as Catherine Redgwell noted in her keynote speech during the 2011 IPCC Expert Meeting on Geoengineering, a "comprehensive assessment of the international regulation of geoengineering," because there does not exist a "single treaty or institution addressing all aspects of geoengineering; rather, the regulatory picture is a diverse and fragmented one both at the international and national levels."[64] None of the documents mentioned were meant to cover large-scale geoengineering experiments like the canceled SPICE test in 2012, or the much-delayed SCoPEx, initially planned for the American Southwest but now being designed for Swedish airspace. When Harvard's SCoPEx Advisory Committee determined that their meso-level study was legally permissible, they considered the Conference of the Parties to the Council for Biological Diversity and its mandate that "no climate-related geo-engineering activities that may affect biodiversity take place, until there is an adequate scientific basis on which to justify such activities and appropriate consideration of the associated risks for the environment."[65] This policy makes exceptions for small-scale scientific research, a guideline to which those involved with SCoPEx believe their work adheres. With no binding national or international agreements that specifically pertain to particle injection testing, however, it is that much more difficult to restrain the myriad subpolitical actors able and willing to perform these kinds of experiments.

PERSUASIVE GEOSOCIAL RISK

At present, numerous subpolitical geoengineering actors have responded to the persuasive, geosocial forces from industrial modernity by focusing on SPI as the most effective and cheapest geoengineering scenario. In emphasizing the potential usefulness of SPI, the geoclique stresses the supposedly low cost of initial implementation, the speed with which it could be made to work, the relative lack of disruption to the world's economies, and the potential for profit. They have, whether intentionally or not, fused SPI geoengineering into Western capitalism, thereby ensuring its appeal to investors. A geosocial understanding of risk allows us to chart three major features of these subpolitical geoengineering arguments. First, many of the research studies developed by the geoclique emphasize "natural" analogs for the risk of SPI, noting the appeal of these supposedly benign or accepted events to generate momentum for research and public policy claims. Second, these groups define the risk of SPI through complex modeling and comparison projects, not unlike the efforts described earlier, thereby enacting a similar kind of discursive-material performance for artificial, atmospheric SO_2 to the kind climate scientists have generated for CO_2. The result of this modeling work has been to create a persuasive state of the future through a speculative, deliberative rhetoric. Finally, these geoclique projects frequently align themselves with ecomodernist principles to rationalize how geoengineering could flourish under economic conditions similar to those that caused runaway climate change in the first place. Through these efforts, SPI is slowly being stabilized as the only viable, long-term geoengineering concept corresponding with known projections of our climate future.

SPI has been a fixture for geoengineering subpolitical actors since Paul Crutzen wrote a brief opinion piece positioning it as a potential fix to uncontrollable or undesirable climate change in 2006. Crutzen accentuated "natural" analogs, including the popular example of the 1991 Mount Pinatubo eruption, and suggested that "sulfate particle concentrations [could] cool the planet, offsetting an uncertain fraction of the anthropogenic increase in greenhouse gas warming."[66] Referencing the half-million annual deaths currently caused by sulfate pollution, Crutzen posited that because of the seriousness of climate change and our apparent inability to act upon it, "the usefulness of artificially enhancing earth's albedo and thereby cooling climate by adding sunlight reflecting aerosol in the stratosphere . . . might again be explored and debated as a way to . . . counteract the climate forcing of growing CO_2 emissions."[67] Crutzen's reputation ensured that the article would garner attention, and SPI was thereby pulled into the modern geoengineering agenda. Yet, his article was not the first to mention the concept in the modern era, an honor that belongs to David Keith and Hadi Dowlatabadi's 1992 examination of geoengineering options. Importantly, this categorical work separated CO_2 reduction techniques from albedo enhancement strategies and delineated different proposed plans for each, articulating their benefits and risks. In terms of the latter, Keith and Dowlatabadi identified what would become the main concerns for SPI investigations. They pointed out that while we have little information to use in a quantitative assessment of risk, we can perform comparative assessments of risk "if a geoengineering scheme imitates or amplifies a natural process," which they believe SPI does.[68] Since the publication of Keith and Dowlatabadi's work, researchers in this vein have stated their preference for climate solutions other than geoengineering interventions, though they continue to frame SPI within the scope of existing natural phenomena—a strategy that we will also see used by geoengineering profiteers in the following chapter.

Multiple research groups have picked up on these initial research suggestions and sought to determine SPI's potential for success by using climate models to determine risk. In these studies, a modeling simulation is set up allowing for the unabated increase in CO_2 for a period of time, say until the year 2100 or until a certain threshold of ppm is achieved. Then, incoming solar radiation is reduced to match expected levels that accord with

SPI geoengineering hypotheses. Matthews and Caldeira's 2007 study, for example, determined what it would take to bring global temperatures down from a 3.5°C increase in 2100, "in response to an atmospheric CO_2 increase from 280 to 880 ppm by volume."[69] Running the model with a "fully geoengineered simulation" in effect, they found global temperatures returned to near 1900-levels, though they also noted a significant risk: global average rainfall decreased by 0.18mm/day, with much less rainfall falling on tropical land areas.[70] Matthews and Caldeira also modeled what would happen if SPI geoengineering were to be abruptly halted. They found that "global temperatures converged rapidly with the nongeoengineered climate," resulting in "several years of very rapid climate change, with annual temperature changes reaching 4°C per decade," or roughly twenty times the current rate of warming.[71] Such details might lend themselves to dramatic headlines, but they also tether descriptions of risk to SPI deliberations. Because this modeling requires vastly complex, hypothetical scenarios, these efforts have been able to generate claims about a future reality for an earth with additional atmospheric SO_2. Currently, only climate modeling studies like these can generate substantive data about the specific risks carried by SPI; essentially, studies like these (similar to previous modeling efforts for CO_2) have performed increased atmospheric SO_2 into existence, and with it, an articulation of risk. Matthews and Caldeira's modeling study is just one of many, and other comparison projects are underway, including the UK-based Integrated Assessment of Geoengineering Proposals and the US-based Geoengineering Model Intercomparison Project, each helping to define risk for SPI and determine what a geoengineered future could look like.

In performing these future states of the SO_2-enhanced atmosphere into a modeled reality, these projects also make a deliberative argument about what future states are possible and what ones carry the most "risk," thereby binding their definitions of risk to SPI. Kravitz et al. have examined these models and argue that none "shows the ability to return both temperature and precipitation to preindustrial levels from a world with high CO_2 concentrations," which means that an implemented SPI will likely result in lower levels of precipitation—something that would *not* be felt equally by all people.[72] Research by the geoclique has connected SPI with a definition of risk—and by controlling and applying models, the model variables themselves have been all but eliminated as the source of uncertainty. We now have a detailed-enough definition of SPI risk that researchers are able to attempt field-based experiments. The Stratospheric Controlled Perturbation Experiment (SCoPEx), a meso-level geoengineering experiment, is perhaps the most likely to be completed in the near future. Led by chemistry researchers Frank Keutch and

David Keith, and under the auspices of Harvard's Solar Geoengineering Research Program, SCoPEx is sponsored by several million dollars from private foundations. The initial designs for the experiment consisted of two controlled balloons that would climb to twenty kilometers and then release a few hundred grams of calcium carbonate—not sulfur dioxide—and then record data about how the small plumes are dispersed in the atmosphere. Keith has defended the project by arguing that "despite all of the concerns, we can't find any areas that would be definitely worse off. . . . If solar geoengineering is as good as what is shown in these models, it would be crazy not to take it seriously."[73]

While Keith considers SPI risks as "undoubtedly substantive," he is also concerned with identifying common ethical norms in the unequal distribution of harm and in identifying "an integrative environmental ethic" to help guide these projects.[74] This integrative environmental ethic is likely to be allied with ecomodernism and is reliant upon the competitive capitalism of industrial modernity that instigated climate change. Along with Ted Nordhaus, Michael Shellenberger, Stewart Brand, and others, Keith is a signatory to "An Ecomodernist Manifesto," a document arguing for the use of advanced technology as a means out of our current climate crisis. This manifesto does not shy away from the logics of capitalism but instead embraces the economic formations that led to destructive fossil fuel extraction and helped to project the geologic risks of industrial modernity onto the present day. In ecomodernism, "more-productive economies are wealthier economies, capable of better meeting human needs while committing more of their economic surplus to non-economic amenities, including better human health, greater human freedom and opportunity, arts, culture, and the conservation of nature."[75] Interested in the creation of a "good" Anthropocene, advocates of ecomodernism pursue private industry funding, especially when state-level backing is thought to move too slowly to address environmental harm. "Accelerated technological progress," the manifesto suggests, "will require the active, assertive, and aggressive participation of private sector entrepreneurs, markets, civil society, and the state," though the state is the final, and least-valued source of this momentum.[76] The Breakthrough Institute, an ecomodern think tank and voice for many in the geoclique, is critical of excessive economic expansion and well aware of capitalism's role in the creation of our current climate crisis; at the same time, it contends that economic models based on degrowth are so unlikely as to be impossible. Capitalism, albeit a green kind of capitalism, is favored as the instrument capable of bringing about climate solutions. As Breakthrough Institute commentator Harry Saunders has argued, we should "not overlook the underlying engine that has driven rising prosperity, slowing population growth rates, and increasing resource efficiency."[77] Generally speaking, ecomodernism is favored by the geoclique as the philosophic and economic strategy that can bring about climate solutions in the Anthropocene.

Perhaps unsurprisingly, Beck has argued that the major tenets of ecomodernism and risk society can functionally coincide. Clarifying how this alignment could occur, and how a new cosmopolitanism might usher in changes to reflexive modernity, Beck imagines that an alternative modernity will need to be built upon existing economic structures, though it should also be able to modify those goals so that "a new version of prosperity" is not defined by unlimited expansion, but by "overall 'wellbeing.'"[78] But this acceptance of capitalism by the geoclique should cause concern. When it is possible to create potential solutions to climate change, patent those technologies, and then conduct experiments outside of the laboratory to determine their side effects, irrespective of international laws, we witness subpolitical actors setting the geoengineering agenda. To such concerns, they respond that they are allowed to pursue these solutions, because governments are moving too slowly to respond to environmental crises and we need to take advantage of an economic system that values innovation. While some actors have gone to great lengths to make their plans transparent and invite feedback on their governance, including Harvard's SCoPEx project, others play by far fewer rules, as their allegiance is to investors, making them less interested in risks that will ultimately be felt by the public.

A useful way to perceive why these sub-political actors embrace capitalism, even a "green capitalism," is by way of Jason Moore's articulation of "Cheap Nature." For Moore, "Cheap Nature" designates how capitalist economies demand that nature be both cheap in price and rendered inferior "in an ethico-political sense, the better to make Nature cheap in price."[79] Moore offers dozens of examples of how capitalist movements have followed this premise, from "sixteenth-century sugar mills, eighteenth-century steam engines, the Fordist assembly lines—all were premised, at every turn, on the appropriation of Cheap Natures. The plantation system was built on Cheap land and labor; steam engines developed at the pitheads of coal mines; the Fordist assembly lines were worthless without Cheap oil, steel, and coal."[80] "But since the 1970s," Moore argues, "the possibilities for securing Cheap Natures have narrowed,"[81] and within that same decade we see the logics of ecomodern geoengineering investment take shape. Capitalism in the Anthropocene, or what Moore calls the Capitalocene, appears to be struggling with how to identify cheap resources in an epoch of scarcity, transformation, and global risk. The abundant natural resources of industrial modernity are now less available and less profitable than before, and, because the effects of industrial modernity's exploitation of Cheap Nature are now being felt, there is more pressure to turn the earth into Cheap Nature and less time to determine how to do so. Though ecomodernism rejects modernity's definition of nature, it seems truly unable to reconceptualize the Western economic relationship with the environment. Caught up in Silicon Valley idealism and techno-utopian visions of "disruption," ecomodern geoengineering approaches depend upon a rationale for Cheap Nature that is still motivated by capitalism. Hence, we have a small network of ecomodern investors, engineers, and scientists working to find environmental solutions that profit in an existing commercial marketplace while simultaneously defining the risks of those global interventions and fusing those risks to commercial solutions. Such a future can never be wholly equitable, because capitalism itself requires unequal access to goods, services, and resources. Geoengineering risk, as it turns out, is a complex matrix of geosocial forces, ones that were set into motion, and continue to be driven, by geologic resource exploitation.

With a geosocial conception of risk, we can more easily locate how capitalism has generated the conditions for risk while also claiming to define it. Geosocial risk is the result of social and economic forces acting upon the earth and the earth issuing forth responses to those same abuses—ones which echo out into social and economic realms. While this risk is a far more slippery concept than any ecoentrepreneur is willing to admit, there are few outside actors to the geoclique who have been able to define or communicate the risks involved. Unfortunately, those who set industry guidelines seem poised to offer nongovernmental regulations that will further support these definitions of risk and encourage further economic development of geoengineering concepts.

The International Organization for Standardization (ISO), whose guidelines for risk management define it in decidedly instrumental terms, also appears to be readying materials for geoengineering implementation. In 2019 the French news service Agence France-Presse revealed a leaked ISO report concerning the draft of a new standard, ISO 14082, "GHG Management—Guidance for the Quantification and Reporting of Radiative Forcing Based Climate Footprint and Mitigation Efforts."[82] While the ISO has clarified that "geoengineering techniques such as Solar Radiation Management and Earth Radiation Management are out of scope of the document,"[83] this standard can only be preparation for SPI geoengineering. Most concerning, however, is that the ISO sets global, industry standards for a range of activities, so the definition for SPI in advance of more widespread practice could be prefiguring risks that are still not well understood. Carroll Muffett, the president of the Centre for International Environmental Law, has argued that "there is a really profound risk when you take something as untested, controversial, politically volatile and morally risky as geoengineering and you make it the subject of industry-driven, market-oriented standards."[84] This risk is heightened when we consider that the ISO is a nongovernmental organization and its members are not delegates of national governments. In practice, then, an approved geoengineering-oriented ISO could give start-ups guidelines with which they can conduct additional, for-profit experiments. As we will see in the following chapter, the marriage of ecomodernism and capitalism can drive some geoengineering subpolitical actors toward the exploitation of vulnerable groups while simultaneously claiming a moral authority.

The risks that we face today are not well served by previous articulations of risk as a scientific fact or as a rhetorical and social construction. Instead, a more accurate definition of risk situates it as a geologic force with social and discursive dimensions that we can identify through the lens of a geologic rhetoric. They are emergent because past environmental abuses are now causing a diffuse, omnipresent, geosocial risk that plays havoc with the present and future. Ulrich Beck's theory of a risk society offers an accurate portrayal of how risk is both diffuse and resultant from Western, industrial fossil fuel extractive practices that stretch back hundreds of years; here, risk is the cumulative result of geologic and social forces. A small group of entrepreneurs and scientists, the geoclique, have taken the reins to act on this diffuse formation of risk, doing so under the banner of ecomodernism and through the technologies of SPI, largely because it is thought to be cheap and fast. Some, like Yuri Izrael, have decided to test their ideas without much governance, while others, like the SCoPEx group, are seeking out additional transparency and feedback. Regardless, these sub-political actors have helped define SPI risk by offering "natural" analogs, by creating climate simulations that double as deliberative future earth states, and by geoengineering experiments that are now being tested in the atmosphere. These subpolitical actors have found that SPI geoengineering can act as a response to climate change and an economy that demands that any response also generate profit. As the following chapter will explore, profit-focused disciplinary thinking is a poor way to pursue any geoengineering goal.

4

ROGUE GEOENGINEERING

The founder of modern geoengineering-oriented science is often said to be Soviet-era scientist Mikhail Budyko. His research, including the article "Heat Balance of the Earth's Surface," has been hailed as a breakthrough moment in the emergence of climatology as a quantitative physical science; in particular Budyko provided one of the first commentaries on how SRM could be employed to manage global temperature.[1] Budyko's research flourished during the Cold War, as US-Soviet hostilities generated the impetus for additional weather management investigations, building from existing military, agricultural, and recreational technologies. During this era, however, few scientists had corresponded the rise of CO_2 to the possibilities inherent in climate-altering technologies even as these connections were made by the president's Science Advisory Committee. That committee authored the 1965 *Restoring the Quality of Our Environment* report, which included details about how rising levels of CO_2 are human-caused and how increased atmospheric CO_2 poses several possible dangers. They incorporated the provocative suggestion that because increased atmospheric CO_2 "could be deleterious from the point of view of human beings . . . the possibilities of deliberately bringing about countervailing climatic changes therefore need to be thoroughly explored."[2] Committee members hypothesized that, if needed, the earth's albedo could be artificially increased through a series of reflective particles distributed onto the world's oceans. While the authors failed to elaborate, they encouraged other researchers to consider the intentional manipulation of climactic CO_2 through engineering efforts.

In this chapter, I explore the emergence of one strain of geoengineering research, the theoretical twentieth-century frameworks that supported it in both lab-based studies and real-world geoengineering attempts, and the development of modern-day entrepreneurial geoengineering that is superficially connected to that research. Looking back to the 1960s and '70s, we can see how some countercultural environmentalists looked to cybernetics theory and devised hypothetical ideas that would eventually become geoengineering concepts. Eventually solidifying into ecomodernism, this line of thought articulated a design methodology that troubled "traditional" environmentalist boundaries between nature and culture, suggesting that there is little difference between the two. These beliefs merged with claims about the process of scientific and technological discovery, or what is also known as techne. For rhetoricians, techne is the art of making and the methods and methodologies through which production occurs. Straightforward scientific methodologies can be identified by an instrumental form of techne, called $techne_1$, where certainty is high and outcomes are predictable, and we can see the history of university-backed ocean fertilization experiments follow this path. However, this $techne_1$ was usurped by cybernetics theories and counterculture environmentalists of the 1960s and '70s, who suggested, in a preemption of the Anthropocene, that human actions have impacted the earth for as long as humans have existed and that there has never been a fundamental distinction between physis and nomos. In this shift, we see an abstract methodology based upon looser relations between physis and nomos. A geologic rhetorical approach reveals how these $techne_2$ methods aim to reconfigure relations between humans and their environment and how they are given free rein to do so by way of cybernetics-influenced ecomodern theories. Geoengineering projects emergent under this framework align their methods with nature when it proves advantageous—in a new manifestation of the "appeal to nature" fallacy. Further, entrepreneurs working with these assumptions are not concerned with fidelity to complex research findings and seem eager to pass off inconclusive methods as investment opportunities. The open-ocean geoengineering "experiment" conducted by Planktos in 2012 is one clear example of this misrepresentation, an exploitation of an Indigenous group, and an indication of how geoengineering could be pursued in the future by other for-profit geoengineering companies.

GEOENGINEERING TECHNE

The impulse to intentionally manipulate the weather has been around for quite some time, but weather manipulation has a relatively distinct methodological goal from that of geoengineering. Weather manipulation is intended to be discreet in its application and effects. For example, cloud-seeding techniques are a long-standing weather control strategy meant to encourage rain in specific locations. In most cloud seeding, silver iodide or potassium iodide is discharged to provide additional particulate formations so clouds can reach a saturation point and thus release precipitation. Many major ski resorts use cloud seeding, especially when ski season demands more fresh powder; it has also been used to good effect to disrupt dangerous hail-producing thunderstorms. Geoengineering, on the other hand, intends to be holistic in its effects. The distinction is similar to that between weather and climate: one is meant to impact a short-term or localized formation and the other a global system. The intention and practice of geoengineering differs from something like fog dispersion at airports; geoengineering is the intentional manipulation of the whole earth system in order to reverse the effect of GHG emissions and/or control temperature.

Early geoengineering research projects aimed at decreasing global CO_2 were fronted by various science luminaries. Widely awarded theoretical physicist Freeman Dyson was among the first to articulate an engineered response to increasing CO_2 in 1977 (though this work was not taken seriously). His assessment of human industrial activity was decidedly pragmatic, and his work assumed that "it seems inevitable that we shall continue for many decades to burn fossil fuels and to increase the level of atmospheric CO_2, without knowing with any degree of certainty the consequences of our actions."[3] Nevertheless, Dyson argued, "in case of a world-wide emergency," we could simply "plant enough trees and other fast-growing plants to absorb the excess CO_2 and bring the annual increase to a halt."[4] Dyson's solution was based entirely on quantitative estimates and the belief that there is "no law of physics or of ecology that would prevent us from taking action to halt or reverse the growth of atmospheric CO_2 within a few years."[5] Using a method similar to Dyson's, Edward Teller, a theoretical physicist and father of the hydrogen bomb, likewise explored human-controlled means of controlling the climate. Teller sought out physics-based solutions to climate change, and his work with Roderick Hyde and Lowell Wood led to the hypothesis that increasing the earth's albedo through SPI or deploying "metallic scattering screens" could assist in lowering the global temperature. They suggested that the problem of CO_2 could be "solved by cooperative application of modern technologies rather than by international measures focused on prohibitions."[6] Both Dyson's and Teller's ideas were grounded upon physics-specific knowledge practices, and while their hypothetical solutions were not meant to be implemented, their studies constituted the first wave of hypothetical geoengineering research.

We can categorize these early geoengineering proposals for removing excess atmospheric CO_2 as narrow, discipline-specific strategies that suggest how one technological concept could be applied to complex earth systems. Solutions that are built from within a single disciplinary framework can feel satisfying because their rationales are so neatly contained. They seemingly provide a one-stop shop for rectifying multifaceted, global problems. Yet, the research methods these scientists employed are an obvious ill-fit to the context of their application, especially as they were intended to be applied within an extremely convoluted thermodynamic system that is more intricate than any one model, method, or field, can currently understand, regardless of the prominence of its origins. This research aligns with a version of techne defined by its limited methods and application.

Techne is a general term for "making" or "craft;" the Greek term also forms the root of modern English "technology." In *Of Art and Wisdom: Plato's Understanding of Techne*, David Roochnik explains that the concept of techne comes from the Indo-European root "tek," which means "to fit together the woodwork of a woven house."[7] In its pre-Homeric meaning, techne "came to refer to the knowledge or skill of the tekton, he who *produces* something from wood."[8] Most rhetoricians are familiar with techne, not because they are also woodworkers but because techne is said to describe the art of rhetoric. Techne involves knowledge about how something is made, the practical issues and goals of that knowledge, the productive skills and abilities that could be employed, and the theoretical rationale for making. As such, techne can be described as "the process of production by which something comes to be" and has historically been understood as a kind of nomos, or the result of human cultural production, as opposed to physis, the energy and internal emergence of the natural world.[9] The historical distinction between techne as nomos, and not physis, will become more important a bit later in this chapter.

Techne can certainly mean different things to different groups, and there exists more than one kind of techne, or making process. Janet Atwill, perhaps the foremost commentator on rhetoric and techne, has argued that techne is never representational of a body of knowledge, because it resists identification with a fixed, normative subject. For Atwill, techne is "knowledge as production, not product, and as intervention and articulation rather than representation."[10] Significant here is that techne is the knowledge *and* practice connected to making and to the materials a particular culture can access, and not just the results of those actions. Techne is never a static form of information but instead a practice that must be engaged for it to be said to be complete. Roochnik argues that there were two major forms of techne in ancient Greece—what he calls a techne_1 and a techne_2. A techne_1 is a precise, determinate, easily recognizable, teachable, and universal process or set of techniques used to accomplish something. This techne might be most closely aligned with a set of direct commands in an instruction manual, recipe, or a linear procedure. A techne_1 concerns how to accomplish something specific, using fairly unambiguous, tested methods—methods that often work within specific situations and under certain assumptions. While Dyson's and Teller's physics-based solutions to excess CO_2 made sense within the narrow world of physics, they were unlikely to be successfully applied to the real world. Their math, we might say, checks out, despite being largely ignorant of the myriad other variables that cannot be contained by their thought experiments. We can call their approach a $techne_1$ method because it was practical and teachable, though it was only practical and teachable within the narrow confines of their disciplines.[11] Their hypothetical methods assumed a relatively straightforward relationship between physis and nomos—one borrowed from a Western tradition that extends back through Greek rhetoric.

$Techne_1$ methods involve interventions into nature through human action and result in products caused by this action, rendering the distinction between techne and physis clear. Without delving into the much larger discussion about the nature of Aristotelian movement or differing Greek schools of thought, we can show this to be generally true. Atwill has explained that Aristotle's "definition of physis is completely dependent upon techne. The mutually constitutive relationship between art and nature is part of the very structure of Aristotelian thought."[12] Such a framework is based on assumptions about how technology and science operate upon the world: humans create technologies, perhaps inspired by nature, and then use those technologies to make use of nature in feats of creation. Techne, the art of making, is therefore aligned with nomos and distinct from physis—that which is used to make or that which is made. Physis, after all, requires no human intervention or ingenuity to continue reproducing, but it can be intentionally manipulated through nomos. This framework is overly simplistic, to be sure, and based on Western, Cartesian, reproducible, field-specific means of production.

When applied as a kind of solution that disregards context, this way of thinking can also fall into a category of ideas that have gone by several names over the years, though most know by the term "technological solutionism." The phrase was first coined by famed nuclear physicist Alvin Weinberg in an essay that asked the question of whether a technology, correctly deployed, could replace the all-too-difficult human work of persuasion. Weinberg, head of the Oak Ridge National Laboratory, lamented that "social problems are much more difficult than are technological problems. A social problem exists because many people behave, individually, in a socially unacceptable way."[13] He was "appalled by the difficulties faced by the social engineer," in part because "motivating or forcing people to behave more rationally . . . is a frustrating business."[14] This is a lament familiar to many modern rhetoricians, even if "rational" action is now seen to be a delusion of the Western philosophical tradition, with little adherence to reality. Nevertheless, Weinberg suggested that we should identify "Quick Technological Fixes for profound and almost infinitely complicated social problems, 'fixes' that are within the grasp of modern technology, and which would either eliminate the original social problem without requiring a change in the individual's social attitudes, or would so alter the problem as to make its resolution more feasible."[15] Initially framed as a *rhetorical strategy*, a technological solution describes a social persuasion problem as technological one and identifies determinate, discipline-specific answers that can be implemented toward a rational result. Technology critic Evgeny Morozov sums up technological solutionism nicely as "an unhealthy preoccupation with sexy, monumental, and narrow-minded solutions—the kind of stuff that wows audiences at TED Conferences—to problems that are extremely complex, fluid, and contentious."[16] This mode of thinking is a powerful motivator for those that believe the real problem with climate change is that technology has not advanced fast enough to avoid it, a view shared by many geoengineering entrepreneurs and those with ecomodern worldviews.[17] Technological solutions are generally representative of a $techne_1$, where rule-oriented, discipline-specific procedures or methods result in a supposedly clear and direct resolution. Dyson's and Teller's geoengineering ideas also fit within this paradigm, as one-stop answers to the problem of global carbon emissions.

However, research papers like those from Dyson and Teller are hardly representative of all geoengineering research; instead, they characterize the first attempts at geoengineering solutions to global environmental problems. While it is exceptionally easy to see hypothetical physics-oriented proposals from Cold War–era scientists as hubristic and ham-fisted, the line of thought from this $techne_1$ technological solutionism continues to the present day. Numerous geoengineering proposals still employ a single technology or narrow disciplinary methodology at the core of their research inquiries. As environmental ethicist Dane Scott explains, geoengineering approaches are often textbook examples of a technological fix.[18] "When the climate problem is reframed as an engineering puzzle involving manipulating stocks of atmospheric carbon (CDR) or the solar budget (SRM)," Scott argues, "crisp, clean, technical solutions seem to emerge."[19] These solutions are admittedly less messy than complex persuasion or wholesale social change, as "instead of promoting change in political and social structures, we are urged to resort to technological solutions that will bypass the blockages."[20] Understanding where these geoengineering versions of technological solutions came from requires us to trace a line of thinking, and its associated techne, from Weinberg's time until today. While modern geoengineering techne has its origins in the thought experiments mentioned previously, it truly took flight in the countercultural movement of the 1960s and '70s, the emergent computing technology companies that that gave rise to the Silicon Valley tech industry, and the cybernetics theories that supported these ideas.

ENVIRONMENTAL DESIGNERS

In the 1960s American countercultural and environmental beliefs converged around San Francisco and the University of California at Berkeley, the location providing fertile ground for activists and freethinkers. Historian Fred Turner argues that the development of progressive politics and the emergence of a broader environmentalist movement in this era fomented convergent thinking between the two, best represented in the ideology of publisher and thinker Stewart Brand. Brand was one of Ken Kesey's Merry Pranksters and the driving force behind the *Whole Earth Catalog*, from which modern homesteaders—whether those living on communes or those simply living "off the grid"—could buy materials useful for the conditions of everyday life. The tagline for the catalog was simply "access to tools," a statement that belied the huge number of technologies available within, including garden implements, clothing, books, maps, camping gear, and (eventually) home computing equipment. Much of this paraphernalia was of interest to those who lived alternative lifestyles and fashioned themselves as part of a "new communalist" movement, the general name for those seeking to live outside the bounds of the traditional family unit and the urban/suburban family house, escaping the supposed disintegration of Western society while doing so. New communalism represented the convergence of several trajectories, including the appearance of a unique leftist counterculture, the rise of environmentalism, concerns over Cold-War nuclear policies, and "white flight" from urban centers. Together, these cultural shifts encouraged some people to move away from cities, focus on self-sustainability, and pursue lifestyles frequently focused on anti-capitalist, antiwar values. Brand's work epitomized this zeitgeist, and the *Whole Earth Catalog* nurtured these cultural desires with consumable goods.

A main function of the catalog was to encourage people to control their own circumstances and choose a more environmentally friendly path in their lives and livelihoods. At Stanford University, Brand took courses with Paul R. Ehrlich, whose bestselling, fear-mongering book *The Population Bomb* forecast a doomsday scenario based on ever-increasing human population.[21] Ehrlich put growing environmental concerns into an identifiable, although alarmist, frame. Subsumed within the cultural reaction to *The Population Bomb*, and earlier books such as Rachel Carson's 1962 *Silent Spring*, was an argument that the earth was growing increasingly unable to support Western abundance and environmental exploitation. Population growth combined with a continual abuse of the earth would result in stressed agricultural systems and fierce competition for natural resources, including arable land. The result of these overworked structures could lead to political tensions, military conflict, widespread pollution, and megafauna collapse. The *Whole Earth Catalog* fed into the belief that people could avoid such catastrophe by placing themselves outside the systems that supported these stress factors, and that they could make this shift with the help of consumable technologies. People were encouraged to see their actions as a response to the state of the world and to purchase the means for taking control—or to put it another way, they were encouraged to design their own circumstances within market capitalism.

Part of a group of technologically savvy environmentalists, Brand was influenced by Buckminster Fuller and his concept of the "Comprehensive Designer"—the notion that one person, or a small group of people, could understand complex systems and design solutions for them. The Comprehensive Designer could "be aware of the system's need for balance and the current deployment of its resources. He would then act as a 'harvester of the potentials of the realm,' gathering up the products and techniques of industry and redistributing them in accord with the systemic patterns that only he and other comprehensivists could perceive."[22] A deeply elitist perspective of design, this theory presumed that a privileged group could grasp spectacularly complex difficulties and devise new concepts to make sense out of possible technological, cultural, and political situations involved, all while remaining rational, practical, and objective. Supposedly, a comprehensive designer "would finally see what the bureaucrat could not: the whole picture," a framing that also underscores Brand's choice of cover photograph for the first catalog—the whole earth.[23] This hubristic assessment echoes throughout the *Whole Earth Catalog*, as the 1969 edition proclaimed: "We are as gods and might as well get good at it. So far, remotely done power and glory—as via government, big business, formal education, church—has succeeded to the point where gross defects obscure actual gains. In response to this dilemma and to these gains a realm of intimate, personal power is developing—the power of the individual to conduct his own education, find his own inspiration, shape his own environment, and share his adventure with whoever is interested. Tools that aid this process are sought and promoted by the WHOLE EARTH CATALOG."[24] Brand's catalog empowered environmentalists to see themselves as comprehensive visionaries with the agency to change their relationship with nature. They were encouraged to understand and control their local relations and reject formal, state-level, or institutional structures that were destroying the environment. Yet these environmentalists were also encouraged to move away from society in order to achieve this rationalists' dream—something that would change as this line of thought moved from encouraging isolationism to encouraging active public engagement.

The version of environmentalism found in Brand's catalog embraced technological advancement for the betterment of the earth and represented a social movement made up of a fusion of "social liberalism and western individualism with a special appreciation for technological know-how."[25] Historian Andrew Kirk argues that in this movement, "the seemingly neat bipolar world of twentieth-century environmental politics became a messy mélange of apparently incongruous philosophies and goals united under the banner of whole systems, cybernetics, and alternative technology," and the design- and engineering-oriented freethinking philosophy of cybernetics influenced this movement most of all.[26] Mathematician and philosopher Norbert Wiener, an early cybernetics commentator, defined it as "the scientific study of control and communication in the animal and the machine," though a somewhat better definition might be gleaned from the etymology of the term.[27] As used by Plato in *The Alcibiades*, the word cybernetics refers to the governance of complex systems, political and otherwise; the word references a "steersman," suggestive of piloting a ship or a group of people.[28] Cybernetics theory offered countercultural freethinkers license to explore the connections between technology, environmentally friendly actions, and comprehensive design.

Especially influential to environmentalists like Brand were Gregory Bateson's extensive cybernetic theories. Bateson's book *Steps to an Ecology of Mind* outlined principles for philosophy, communication, psychology, and environmentalism, and thus solved a key dilemma for many new communalists who also appreciated the ability and power of the comprehensive designer.[29] That is, while living "off grid" could be a means to design your own closed system, it also avoided participation in the larger social system and thus neglected an individual's social responsibility. Bateson's resolution was to claim that all phenomena in the universe are connected by cause-and-effect chains of action and reaction, systems he identified as "closed." However, Bateson also argued that these same systems, "notably organisms in environments, ecosystems, thermostats, steam engines with governors, societies, computers, and the like," were also "open" in that they could (and sometimes should) be influenced by outside actors.[30] When applied to human relationships with the environment, this insight suggested that humans operate in a global ecosystem that is sustained as a series of perpetuating closed systems; at the same time, people also need to function within those systems in order to change them. Bateson reasoned that if we ignore how the earth system works and disregard our responsibilities to affect positive change, we are ultimately doomed to fail in our endeavors. If the larger culture in which people exist continues to act as though nature were ours to exploit, Bateson reasoned, the "likelihood of survival will be that of a snowball in hell. You will die either of the toxic by-products of your own hate, or, simply, of over-population and overgrazing. The raw materials of the world are finite."[31] While some found a common-sense environmentalism in these claims, others found permission to envision changing the whole earth system by technological means.

Bateson's environmentalist thought is in some ways a precursor to the naming of the Anthropocene. In Bateson's theory, humans are agents for change that work in coordination with the natural world, but he also considered human consciousness as part of a global ecosystem. "By recognizing the degree of their integration into the natural and social systems around them," Bateson claimed that "individuals could simultaneously restore their individual humanity and act more humanely toward the planet as a whole."[32] In this configuration, humans are always already of the world system and have no need to (literally or philosophically) remove themselves from the larger communities and environments of which they are a part; "they could simply work to influence whatever local 'system' in which they found themselves involved."[33] Those who wanted to function within existing ecological, political, and economic organizations now had a reason to shrug off their new communalist, back-to-the-land inclinations and instead influence how the system worked, with the larger goal of saving the planet. Bateson warned that "nobody knows how long we have, under the present system, before some disaster strikes us, more serious than the destruction of any group of nations. The most important task today is, perhaps, to learn to think in the new way."[34] This "new way" became the integration of ecological frameworks, cybernetics ideals, and environmental thinking into existing capitalist economics, so that environmentalism would become diffused throughout Western politics. This change—from an environmentalism that avoided interactions with culture to one that actively sought out cultural intervention—precipitated the merging of a new era in environmental thought that incorporated cybernetics philosophies with the impulse to work within existing systems. As a result, this shift altered some of the basic components assumed in a $techne_1$, as physis and nomos were no longer considered wholly distinct. Already part of the complex natural world, humans and their actions—what were classically thought of as nomos—were merged with the natural—physis. Wildly heterodox solutions to environmental problems were seen not as simply technological solutions but as complex interventions into the Whole Earth.

ECOMODERNIST TECHNE

By 1974 Brand had used some of the profits from the *Whole Earth Catalog* to found *CoEvolution Quarterly* (CQ), a magazine for broader inquiries on environmental and philosophical thinking than the catalog allowed. Soon after, readers were introduced to Gerard O'Neill, a physics professor at Princeton University, and the idea of "space activism," the advancement of space exploration and colonization. Hypothesizing on what would become known later as "O'Neill Cylinders," he argued that if global environmental conditions became bad enough, humans could live comfortably in space in two massive counter-rotating cylinders that would mimic gravity through centrifugal force. What might have been assumed to be a harmless thought experiment sparked a substantial controversy; the article generated so many angry letters that Brand opened up a section of the following issue to give voice to the debate. For environmentalists, "the space program, let alone space colonies, represented a dangerous distraction from the very immediate environmental problems on earth."[35] Instead of advocating for reducing or reversing environmental damage, O'Neill suggested that we could pursue a solution that abandoned (what were thought to be) established environmental principles. The cylinders were an attempt to decouple human life from the sustained well-being of the planet and pitted those interested in a vision of environmental activism, degrowth, and sustainability against those focused on large-scale solutions intended to "buy time" to solve environmental problems.

By making the methods and logic of this solution public, O'Neill, Brand, and *CQ* magazine offered these radical ideas up to criticism, thereby exposing the tacit assumptions of the Western environmental movement. Counterculture environmentalists like Brand were after exactly this kind of social critique—one that would both shock people into recognizing the severity of our global problems and prompt them into action. Brand may never have believed that the O'Neill cylinders should be pursued, but they created a heated discussion about the future of the environment. Far from a hypothesis dependent upon one disciplinary field, like Dyson's and Teller's work, O'Neill's ideas drew upon multiple fields, considered the methods, logistics, and costs of implementation and conjectured about the forms of societal change that could accommodate, or be altered by, the application of these admittedly far-fetched methods. The O'Neill Cylinders were still clearly the ideas of someone hoping to be seen as a comprehensive designer, but they were also an attempt to move beyond an instrumental $techne_1$ plan represented by only a single discipline and are thus representative of a more complex methodology.[36] Recall that, according to Roochnik, a techne$_2$ represents a more "messy, less determinate, more personally engaging affair than that of techne$_1$."[37] This form of techne does not have a "fixed or invariable subject matter" and is not infallibly certifiable, teachable, or reliable.[38] That is, techne$_2$ kinds of activities are not based upon a supposedly clear-cut set of steps but instead operate in a world with less definitive knowledge and greater variables. Though still a kind of thought experiment, the techne that the O'Neill Cylinders represented fits this description and helped to identify the "shifting and contestable borders of what is possible," a techne not confined to the limitations of existing knowledge, no matter how secure or certain they are.[39]

The emergence of a techne$_2$ for environmentalist thought not only marked the borders of what was possible, it also opened the door for a different strain of research that would lead to current-day geoengineering. In identifying five different definitions for techne, Kelly Pender sees complexity beyond the most instrumental forms and explains the broader social effects of less-determinate techne. Pender describes how this more involved, abstract techne can be defined as "a means of inventing new social possibilities."[40] Certainly, the O'Neill Cylinders were intended to do just that by forcing environmental thinkers to consider new configurations of social and environmental life. Further, while a more determinate techne$_1$ relies upon a stable definition of nature, a variable and complex techne$_2$ is not based upon such a clear foundation. Ancient Greek techne was intended to imitate nature, and any application of a techne was supposedly meant to advance in a manner similar to natural processes, but this connection is removed in more abstract forms of techne. Atwill's work is again insightful here. As part of her larger approach to identify a techne outside of a liberal humanist model, she notes how some techne is never static, how it resists identification with a normative subject, and how it includes the potential to transgress existing boundaries to redefine the relations of power.[41] While Atwill emphasizes the role of exchange between maker and user, and focuses on techne as mediating relations between people, we can also envision techne as the assembling and re-organizing of human and nonhuman actors in the service of a larger goal. That is, a more abstract techne can redefine existing power structures, and can potentially reconfigure the relations between nature and culture.

A techne$_2$ does not rely on a stable, fixed meaning for physis, like techne in a strict Aristotelian sense. In Aristotle's conception, a "definition of physis is completely dependent upon techne," where physis is supported by an underlying, primary substance.[42] The clarity of this relationship is expressed by techne$_1$, but a techne$_2$ does not have a fixed substance, or a well-defined version of physis, upon which it rests. For the Greeks, it would have been unthinkable for the maker (technite) to be removed from their cultural context or the ethics of designing for their social environment. A technite who was disassociated from their culture would not have been a master craftsperson but only a "'copyist,' performing blind acts without knowing why."[43] However, the immediacy of this closed system is no longer applicable for environmental positions that break with traditional, Western understandings of how humans are connected to nature. As the O'Neill Cylinders dramatically showcased, while older horizontalist $techne_1$ methods might dictate how humans could intervene into a known and stable world, newer strategies could reject any clear distinction between nature and culture. Here, an abstract $techne_2$ method could claim to reconfigure the relations between humans and their environment—to alter physis and nomos.

Brand's efforts to popularize an environmentalism based on the radical potential of humanity, combined with the prospect of these new $techne_2$ hypotheticals, emerged into what has been called "ecomodernism," "ecopragmatism," or "postenvironmentalism." Ecomodernism claims to be based upon the Enlightenment ideals of progress and rationality and the concept that modern environmental action "is possible only on the condition that we embrace human development, modernization, and technological innovation."[44] Ted Nordhaus and Michael Shellenberger's 2005 essay "The Death of Environmentalism: Global Warming Politics in a Post-Environmental World" extended the discussion begun by cybernetics theorists and Brand's publishing. Nordhaus and Shellenberger argued that in order to meet the challenges of the future, environmental thinking would have to change, because "modern environmentalism is no longer capable of dealing with the world's most serious ecological crisis"—that of global warming.[45] They take (what they call) "reversionary" environmentalists to task, accusing environmental policy makers of defending a special-interest area of politics and holding fast to a delineation of nature and culture that can no longer be sustained. "Most environmentalists don't think of 'the environment' as a mental category at all," Nordhaus and Shellenberger explain, "they think of it as a real 'thing' to be protected and defended."[46] For ecomodernists, the policies based on the assumption of a stable distinction between physis and nomos are neither productive in a world of increasing environmental and economic pressures nor realistic in the Anthropocene—and their approaches to environmental problems mirror this belief.

Ecomodernists claim that the only constructive way forward is to think beyond long-standing categories of environmentalist thought like conservationism, preservationism, or land-ethics and move into a new era of environmental strategies carefully managed to create a "good Anthropocene." As described in the previous chapter, ecomodern commentators claim that this Anthropocene should be "designed and planned by humans using their growing social, economic, and technological powers to make life better for people, stabilize the climate, and protect the natural world."[47] The authors of the aforementioned "Ecomodernist Manifesto" contend that any fundamental change in the Anthropocene must necessarily recognize that human activity has long been an essential driver of environmental change. While some of these changes have been good, there have also been many negative ones, like the discharge of pollutants, the loss of the ozone layer, and the release of CO_2 into the atmosphere. For ecomodernists, "meaningful climate mitigation is fundamentally a technological challenge. . . . Absent profound technological change there is no credible path to meaningful climate mitigation."[48] Decidedly the legacy of cybernetic-influenced environmental thinking, ecomodernism posits that the future of the environment, and human habitation of the earth, rests on the shoulders of those willing (and able) to provide technological stewardship. We are as gods, as the argument goes, so we might as well get good at acting that way.

The problem many have with this position, however, is that it simultaneously contends that technology will save us and denies that there is such a thing as "the environment," out there, beyond human culture. As such, it blurs the line as to what a technological intervention, and therefore what techne, actually is. Here, an ecomodern techne is defined as the human art of creating new materials and implementing new techniques, but a human art that is always interlocked with a long history of human interaction with and dependence upon nature. In an Anthropocene epoch, where nature is shot through with human influence, and where human action is always natural, there is little value in defining physis and nomos as separate. Ecomodernism, for all its claims about the value of technological intervention, generally fails to distinguish between nature and culture, and assuming that there is a difference is to perpetuate a belief that is out-of-step with our Anthropocene reality. This distinction is precisely where the ecomodernist viewpoint on techne differs from other environmental positions, insofar as ecomodernism simultaneously embraces the power of a designed intervention, recognizes the complexity of that intervention, and sees techne as able to mediate these abstract relationships. Instead of trying to police a boundary as to where nature resides, ecomodernist methods of design assume that every method is grounded in nature—even if we are unable to easily recognize or articulate how. This matrix of slippery definitions has permitted geoengineering subpolitical actors to play fast and loose with scientific methodologies, arguing for the "natural" features of their proposed solutions and thereby committing a highly specious "appeal to nature." Further, these subpolitical actors can also exploit marginalized populations in order to generate what is ultimately intended to be a consumable method or product, all while testing those solutions amid ambiguous legal contexts—legal contexts established by environmental legislation based upon more traditional definitions of physis and nomos. The following example about open-ocean iron fertilization experiments showcases how $techne_1$ operated and how a $techne_2$ approach deviated from the scientific research when placed in the service of profit-seeking geoengineering entrepreneurs. It serves, I argue, as a warning as to how geoengineering could be pursued in the near future.

THE IRON HYPOTHESIS

In the summer of 2012, some three hundred miles off the coast of the Canadian archipelago known as Haida Gwaii (X̱aayda gwaay), a surprisingly large-scale oceanography experiment was carried out. Throughout July and August, Planktos Inc., working with (and possibly *as*) the Haida Salmon Restoration Corporation (HSRC), intentionally poured 120 metric tons of iron sulfate and iron oxide into the Pacific Ocean, "seeding" more than 3,800 square miles. This form of geoengineering fits under the broader label of CDR and seeks to achieve this goal through the intentional manipulation of aquatic life. By dumping iron filings into the ocean, these ocean-fertilization strategies attempt to increase the phytoplankton biomass, because phytoplankton thrive on naturally occurring iron. When additional iron is added to the ocean, phytoplankton overproduce and bloom in larger numbers, triggering zooplankton to eat the phytoplankton. Ideally, this increased biomass pulls more CO_2 out of the atmosphere, and when those phytoplankton die, they (hopefully) sink, depositing CO_2 on the bottom of the ocean. Some previous iron dumping tests suggested that an increase of this sort could possibly sequester CO_2, though these claims are now largely denied by the scientific community. As one might imagine, the intentional dumping of such a large amount of iron into the ocean was met with widespread criticism. The legal ramifications of Planktos's actions are not straightforward, but this experiment may have been in violation of the Convention on the Prevention of Marine Pollution by Dumping Wastes and Other Matter (1972), the Convention on Biological Diversity (1992), the Canadian Environmental Protection Act (1999), and the Cartagena Protocol of Convention on Biodiversity (2000).[49] But, like with so many geoengineering proposals, none of these agreements was designed to fit contemporary circumstances.

The Planktos "experiment" also received a significant response from marine scientists and environmental organizations who were outraged that a private company was allowed to dump foreign material into the ocean, with no apparent restrictions or oversight. From Planktos's perspective, however, the experiment had environmental, humanitarian, and economic rationales. Planktos claimed they used materials that were already "natural" to the ocean, only enhancing the concentration and volume of iron in one location (something that occurs from volcanic activity). Complicating the issue was the initial cooperation from some Haida Indigenous community members, who were interested in whether iron dumping would increase salmon stocks and curious as to whether such a partnership with a rogue geoengineering company could bolster their economic and cultural independence. Yet, the larger problems with this kind of geoengineering experiment lie not only with the ignorance of existing international environmental agreements, but also with the processes, methods, and inventional strategies used to develop and implement them. To put it another way, the issue is one of techne. Bolstered by an emergent ecomodernism and suggestions about the "naturalness" of all techne, Planktos formed the Haida Salmon Restoration Corporation, an AstroTurf group run by the CEO of Planktos, as a means to align their work with the Haida's Indigeneity. This move was clearly an attempt to stave off public criticisms and associate its geoengineering techne with the myth of "authentic" ecological nobility that equates Indigeneity to nature. The path that led to Planktos's geoengineering experiment, however, takes us through years of ocean fertilization research and the willful misinterpretation of that work for potential short-term profits.

The group of geoengineering ideas described by some as "ocean fertilization" has been thoroughly studied by marine biologists seeking to understand the relationship between naturally occurring iron deposits in the world's oceans and the plants that thrive on them. Micronutrient iron occurs in the ocean as a result of geologic forces, and one of the world's largest biomass contingents, phytoplankton, requires this micronutrient iron in order to thrive—similar to how humans require different dietary minerals. The difference is that the presence of abundant micronutrient iron has long been thought to result in large-scale phytoplankton blooms. If an iron-rich event, like volcanic activity, triggers a phytoplankton bloom, it is believed that additional atmospheric CO_2 could be absorbed by the increased number of biotic organisms present. By converting dissolved CO_2, phytoplankton create a deficit in the surface level of the ocean that is compensated for by CO_2 being drawn down from the atmosphere. This potential for CO_2 absorption has been of interest to geoengineering researchers, as some believe it is possible to provoke a phytoplankton bloom through the intentional addition of iron, often in the form of iron sulfate, into waters that have are currently low in micronutrient iron.

While Dyson's and Teller's physics-based geoengineering ideas boasted that any increased atmospheric CO_2 levels could be accounted for and reversed by installing a giant metallic screen in the upper atmosphere, a similarly audacious suggestion concerning a wild techno-fix occurred in ocean fertilization circles. During a 1988 lecture at the Woods Hole Oceanographic Institution, oceanographer John Martin declared, "Give me a half tanker of iron, and I will give you an ice age."[50] Martin knew how such a statement would sound to the assembled scientists and reportedly delivered the line with an accent meant to evoke Dr. Strangelove—the fictional scientist played by Peter Sellers in Stanley Kubrick's movie. Martin was certainly aware, as were the rest of the relevant scientific community, that such claims about the role of micronutrient iron in the world's oceans were premature, as little research had been done to prove their accuracy. In the following years, the scientific community would first concern themselves with a different, preliminary question more appropriate for field-specific scientific inquiry: whether additional iron added to surface areas of the ocean could increase phytoplankton numbers. While not naive to the question about CO_2 sequestration, this "iron hypothesis" hinged on whether phytoplankton blooms were attainable in a controlled setting.

Studies looking to test the iron hypothesis should not be described as attempting a technological fix, nor as geoengineering research, because they were not trying to conclude whether additional iron sulfate would export CO_2 and therefore replace other forms of social change. Some of the first studies about the iron hypothesis came from Martin's own work with Stephen Fitzwater. They determined that previous glacial periods had increased levels of iron (Fe) dust and found that "Fe availability may be important in determining global atmospheric CO_2 levels, and hence, affecting global climate."[51] Martin and Fitzwater's follow-up research with Michael Gordon further established that Fe deficiency in particular high-nutrient, low-chlorophyll zones (like the Gulf of Alaska, the northeast Atlantic, and the Southern Ocean) were directly related to a lack of large phytoplankton blooms.[52] They performed small-scale enrichment experiments in the Ross Sea (a deep bay off of Antarctica) "to test the hypothesis that iron deficiency is responsible for the phytoplankton's failure to use up the luxuriant major nutrient supplies found in these and all other offshore Antarctic ocean waters."[53] These studies also intimated that if Fe determined biomass productivity, then there could exist a clear relationship between Fe availability and CO_2 sequestration.

Much of the scientific research in this era was micro-level in scale and best described by Coale et al. as "bottle enrichment experiments" performed under controlled laboratory conditions by biogeochemists, marine ecologists, and oceanographers.[54] Strong et al. portrayed these conditions as "ultra-clean experimental approaches with seawater incubated on the deck of a ship."[55] This science fits neatly within a $techne_1$ approach, insofar as researchers sought out specific answers while using precise, tested disciplinary methods. By no means is this description negative—much groundwork in science relies upon these kinds of determinate methods. (This method also differed from Dyson's or Teller's work because it was achievable in a laboratory setting.) Later, a new wave of open-ocean, mesolevel experimentation began, with researchers employing much larger tests and measurement methods to determine the real-world effects of iron additions. The next decade was dominated by experiments that intentionally added iron, and eventually iron sulfate ($FeSO_4$), to the ocean. IronEx I (1993) in the Eastern equatorial Pacific Ocean was the first, dispersing 450 kg of iron in an area of sixty-four square kilometers. IronEx II (1995) repeated some of the same methods, at the same volume, but dispersed iron over the course of a week. The results were convincing enough to suggest that the iron hypothesis had been resolved and should now be treated as the "iron theory."[56] That is, studies showed clear causal relationships between iron additions and phytoplankton blooms in a controlled setting but did not show a corresponding relationship between iron and CO_2 sequestration. The authors of the results for IronEx II suggested that further research on the relationship between iron and CO_2 levels was required but cautioned that "owing to extreme turbulence and temporal variability, a mesoscale enrichment experiment in these Southern Ocean waters poses a tantalizing, yet formidable, challenge."[57] The following decade of mesoscale iron enrichment experiments would answer this challenge, broaden the data collection, and ask whether iron fertilization could impact CO_2 levels in the uppermost regions of the Southern Ocean and subarctic Pacific high-nutrient, low-chlorophyll (HNLC) regions.

Between 1999 and 2004, scientific teams conducted nine more mesoscale studies, using between 1,740 kg to 7,000 kg of Fe each and lasting in duration from thirteen to thirty-five days. On the surface, these experiments were still concerned with nuances in the iron hypothesis, determining under what conditions and through which methods iron could induce a phytoplankton bloom, but they also measured—or attempted to measure—CO_2 export to the deep ocean. Strong et al. note that the scientific language in these studies began to change, as inquiries into "carbon export" explicitly shifted to that of "carbon sequestration."[58] This terminology, used by the Department of Energy, pointed to a different end-goal for iron fertilization experimentation and "contributed to blurring the lines between the basic and applied science dimensions" of ocean fertilization.[59] These studies, which go by a disorienting series of acronyms,[60] established the iron theory beyond reasonable doubt but also revealed that the volume of CO_2 transported to the deeper ocean was very modest, leading Buesseler and Boyd (2003) to summarize that a million times the volume of the iron dumped during two of the largest experiments would be needed to offset just 30 percent of the annual carbon output of human activities (a conclusion affirmed by de Baar et al., among others).[61] Researchers appeared to resolve not just the iron hypothesis but also the carbon sequestration question—that is, it was highly unlikely that meaningful carbon sequestration could be achieved through "ocean fertilization" methods. These disciplinary outcomes, however, failed to discourage some entrepreneurs looking to profit from more abstract experimental methods.

OCEAN FERTILIZATION PROFITEERING

By the late 2000s scientists were mostly finished performing open-ocean mesoscale iron research, with the results that the main pieces of the iron hypothesis had been resolved and CO_2 sequestration had been largely rejected. Researchers found no substantive evidence for significant carbon export and found few reasons to continue pursuing this line of investigation. Scientists also issued several warnings as they became increasingly aware that their findings were being investigated by for-profit companies seeking to sell carbon-offset credits and/or patent a method for CO_2 sequestration. Chisholm et al. sounded one of the first alarms, announcing that carbon export had not been verified and that the "consequences and uncertainties of ocean fertilization already far outweigh hypothetical benefits." In surveying existing models, they found that any sustained iron fertilization effort could result in deep ocean hypoxia (low oxygen conditions) or anoxia ("dead zones").[62] Reporting details from the SOFeX study, Wingenter et al. cautioned against any anthropogenic CO_2 sequestration efforts, and Gilbert et al. likewise urged restraint due to possible direct and indirect ecological impacts.[63] Cullen and Boyd's response and Strong et al.'s commentary were also intended to shut down any suggestion that iron fertilization was possible or profitable.[64] Both pointed to the uncertainty of a phytoplankton bloom in response to iron fertilization, the huge number of unknown impacts upon marine ecosystems, and the potential consequences of intentionally changing the levels of carbon and oxygen in the ocean. The final mesoscale iron fertilization experiment is perhaps most emblematic of a change in research trajectories. The 2009 LOHAFEX study planned to distribute ten metric tons of $FeSO_4$ over forty days and $300km^2$. A joint project between the Indian Council of Scientific

Industrial Research and the German Helmholtz Foundation, the expedition was canceled due to claims the study would breach the 2008 Convention on Biological Diversity. LOHAFEX also attracted more attention than previous experiments because of a growing awareness that for-profit companies could capitalize on iron fertilization research and usher in a new approach to geoengineering.

For almost as long as the iron hypothesis has been around, various entrepreneurs have tried to profit from the logical extension of this scientific knowledge—that if the iron hypothesis is true, so too must be assumptions of CO_2 sequestration. Some companies have tried to sell carbon-offset credits to industries who put an abundance of CO_2 into the atmosphere, while others have sought to patent a method for introducing iron sulfate into the ocean by demonstrating what a decade of careful mesoscale research had not found—significant carbon export to the deep ocean. This form of speculation began in 1994 with entrepreneur Mike Markels's patent application for an ocean fertilization method (supposedly) intended to improve seafood production. In 1998 Markels's company, Ocean Farming, conducted two small-scale experiments in the Gulf of Mexico; Markels would later partner with Richard Barber, one of the researchers from the IronEx experiments, to form GreenSea Venture, a company intending to sell carbon-offset credits for the CO_2 they planned to sequester.[65] GreenSea was founded upon specious evidence, with Markels and Barber citing other open-ocean mesoscale experiments, "although none of these experiments showed any carbon export from an iron-induced bloom."[66]

The number of for-profit iron fertilization geoengineering companies grew quickly, however, with other start-ups band wagoning on to GreenSea's optimistic speculation, including Carboncorp USA, Climos, Oceaneos, the Ocean Nourishment Corporation, and Planktos. It would be relatively easy to suggest these companies are interested in technological get-rich-quick fixes. Their profiles, for the most part, promise a return on investment, either through carbon-offset shares or through the longer-term process of patenting a method to sequester carbon. In a *WIRED* magazine interview from 2000, Markels unironically suggested that "given 200 boats, 8.1 million tons of iron, and 16 million square miles of HNLC ocean . . . his flotilla could zero out 8 gigatons of CO_2 each year, the entire global fossil-fuel emissions enchilada, all for an annual cost of around $16 billion. Crisis over."[67] Most experienced researchers would vociferously disagree with Markels's assessment, but expert disagreement has rarely dissuaded entrepreneurial geoengineers from testing the waters.

For-profit ocean fertilization ventures have intentionally misrepresented or willfully misunderstood the complexity of the systems from which they hope to generate revenue; because of this, they have misapplied a methodology for working in that context. They deliberately mistake complex $techne_2$ uncertainty for relatively simple $techne_1$ methods, characterizing the modest evidence shown by mesoscale research as a bankable certainty. Iron fertilization geoengineering companies present themselves as focused on sustainable, "natural," low-cost solutions for repairing the earth's balance of CO_2, while simultaneously marketing them as a means to increase fish stocks. For example, the Ocean Nourishment Corporation asserts they have performed feasibility studies for iron dumping that could alleviate concerns about overfished waters, atmospheric CO_2, and ocean acidification. They also purport that one of their sites removes "5–8 million tonnes of CO_2 from the atmosphere for each year of operation, equivalent to offsetting annual emissions from a typical 1,000 MW coal-fired power station or the short-term sequestration from one million hectares of new growth forest."[68] Such declarations contradict the numerous position papers arguing that iron fertilization is an unsafe, ineffective means of CO_2 sequestration, which moreover shows no direct evidence of facilitating fish stock recovery.[69] Yet, Oceaneos appears to have dropped their focus on selling carbon-offset credits, and now states their iron dumping efforts will bring back fish populations. In order to make their endeavors seem like good investments, an uncertain process with no confirmed scientific support is represented as a determinate process with certain results. Perhaps it is not surprising that for-profit companies are trying to monetize this research and represent iron fertilization, and geoengineering in general, as a knowable, easy, safe, and "natural" method of CO_2 sequestration and ecosystem restoration.

The argument that geoengineering is "natural" takes advantage of claims about the lack of distinction between nature and culture, an argument supported by cybernetics and ecomodernist thinkers. Geoengineering companies look to exploit this confusion, mostly by emphasizing that whatever their geoengineering plan is, it has a natural analog that matches up with modernity's clear distinctions between physis and nomos. By eschewing representations of their work as a human-made techno-fix, and instead framing it as an enhancement of existing earth processes, these companies hope to build public support—or at least generate investment—for their methods. Science communication researchers have already identified the supposed advantages in presenting geoengineering as "natural." In their analysis of the main metaphors of geoengineering, Luokkanen et al. have suggested that by aligning their methods with natural processes, geoengineers might attempt to make their work seem less dangerous and that connections to "naturalness" "may come with a license for free experimenting."[70] In this vein, Markels has argued that his projects focus on existing connections with the natural world and that his work simply nudges nature along.[71] This relationship with nature is only emphasized by popular press books like David Biello's *The Unnatural World*, which portrays bio-oceanographer Victor Smetacek's ocean research as a natural replacement for years of whale hunting or Levitt and Dubner's *SuperFreakonomics*, which even connects Ken Caldeira's name to that of a volcano caldera, offering a particularly sloppy piece of evidence that he, and therefore his research, is aligned to the natural world.[72]

More to the point, the Planktos company connects iron dumping to natural processes by highlighting whales' role as nature's geoengineers, because of their ability to export carbon—they poop it out. Planktos describes their own methods as simple and innovative and their process of "healing the seas" as akin to the natural geologic distribution of micronutrients, relying heavily on evidence from "naturally stimulated iron replenishment events in the Pacific Northwest due to volcanic dust."[73] The argument that Planktos and the duplicitously named HSRC have made depends upon a direct and concerted effort to show that their process of iron fertilization replicates existing natural activities—physis—and that ocean fertilization is in fact simple and straightforward. The actions that Planktos undertook in Haida Gwaii, however, help us locate how geoengineering entrepreneurs intentionally misrepresent scientific methods as a means to persuade others of the economic and environmental potential of their for-profit endeavors.

PLANKTOS AND HAIDA GWAII

Planktos performed their first open-ocean fertilization demonstration in 2002, by pouring iron and paint pigment into the ocean off the coast of Hawaii, using a boat owned by musician Neil Young. Even though he presented no data to back up his claims, Planktos's CEO, Russ George, argued that this experiment "made the city of Half Moon Bay [where Planktos was headquartered] carbon-neutral for a year."[74] Planktos sold carbon "green tags" at $4/ton of CO_2 to finance some of the trip, though the credits were mostly a public relations ploy. Emboldened by the event, Planktos then prepared for a much larger demonstration and in 2007 announced plans to fertilize a 10,000 km^2 patch of ocean 350 miles west of the Galapagos Islands. Greenpeace, the World Wildlife Fund, the US EPA, the Action Group on Erosion, Technology and Concentration (ETC), the Sea Shepherds, and the Conference of Parties to the London Convention, among other actors, forced the project to a halt. (Similar public criticism would later stop the 2009 LOHAFEX study named earlier.) Undaunted, Planktos would try again, this time successfully completing the largest iron fertilization test to date.

Due to the negative publicity around the Galapagos incident, investors left Planktos Inc., which then suspended operations, but the company re-emerged as Planktos Science, and later Planktos Ecosystems. Whatever the name, the company's plans in 2012 were bigger than before. That year, they poured of over 120 metric tons of iron sulfate particles (technically, 100 metric tons of iron sulfate, and 20 metric tons of iron oxide) into the ocean, roughly three hundred miles off the coast of Haida Gwaii in Canada.[75] The town of Old Massett helped to facilitate the project with CAD$2.5 million of their own funds. While Planktos was concerned with generating evidence of CO_2 sequestration and looking to align itself with a more creditable partner, the Old Massett council was "told they would earn carbon credits from the dump" and were hoping for increased salmon stocks.[76] Plankton feed on iron, salmon feed on plankton, and humans feed on salmon. But because of human-caused global warming, long-standing wind patterns that had reliably transported iron into the waters surrounding Haida Gwaii had been on the decline, as had salmon populations and fishing yields.

A main point of concern with Planktos's geoengineering experiment is whether the test constituted a mutually agreed-upon and mutually beneficial exploration for both Planktos/HSRC (they are difficult to separate) and the Haida people. That is, the test was allegedly done in order to demonstrate the scientific and economic potential for iron fertilization, but judging by the lack of evidence for the efficacy of iron fertilization and the presumed lack of an attempt to gather such evidence, that seems an implausible goal. A business based on the iron fertilization model would need to misrepresent the science behind its claims in order to project confidence in its process and product. It appears more likely that Planktos distorted existing evidence about the effects of ocean fertilization research to convince some members of the Old Massett Haida into funding the project. If that is the case, Planktos misrepresented CO_2 sequestration as a reproducible, instrumental technological solution, when in fact only the iron hypothesis seemed to be conclusive. Backed by ecomodern ambivalence about physis and nomos, the Planktos geoengineering project deployed several rhetorical strategies to carry off this deception, ones that also made claims about the "naturalness" of their techne.

The geographic location of the experiment was certainly a key factor in the Haida Gwaii geoengineering event, as Planktos used the history of colonially managed geologic energy to their advantage. Significantly playing a role in this experiment is a colonial history that restricted the amount of accessible, stored carbon energy on the islands. We can trace this deficit to Western contact in the eighteenth century, the widespread transmission of smallpox, the eventual deaths of 95 percent of the Haida people during the nineteenth century, various assimilation practices, and long-standing infrastructural neglect by British and then Canadian governments. The impact of this trauma continues to influence the Haida and other Indigenous groups, who are still "marginalised by nearly every measure of social and economic inclusion."[77] More recently, global warming has amplified ocean temperatures and acidification and has led to the decrease of salmon stocks worldwide. This is especially apparent in the Pacific Northwest, whose overfished waters and declining marine populations are under increasing pressure. In previous cycles of global micronutrients, the waters off Haida Gwaii had been seeded by wind patterns that distributed micronutrient iron sourced by volcanic activity elsewhere in the Pacific, a material also known as Aeolian dust. Those involved in Planktos certainly knew of this geologic history, and while they used the effects of global warming to emphasize how their technologies could help remediate the situation, they also relied on the effects of colonial abuse and postcolonial mistreatment as a reason why the Haida might be interested in the investment opportunity.[78]

The persuasive strategies mentioned here also adjoin conceptions of Haida identity. For the Haida, the resources of the land, including salmon for food and timber for building materials, are forces of deep material and cultural significance. Salmon are a vital part of Haida selfhood, so the desire to support declining fish populations is understandably high, especially since salmon fishing has been put under increasing stress by Canadian regulations.[79] Because Planktos is not connected with the Canadian or US government and is not a scientific organization in the traditional sense, and because geoengineering still exists in a quasi-legal state, the test could have been seen as a step toward independence from larger governmental oversight. The Haida, like many First Nation and Native American groups, are actively engaged in Indigenous sovereignty movements, where self-sufficiency and distinct cultural identity from nation-states are seen as priorities. The Planktos experiment, if successful, could have assisted the Haida in increasing their autonomy from the Canadian government and possibly bolster their ongoing "geologic selfwitnessing" and independence.[80] Planktos offered the possibility of restoring the Haida tradition of harvesting halibut and Dungeness crabs in the spring and sockeye, spring, coho, and pink salmon in the summer.[81] While some might see the geoengineering test as a threat to nature, for some Haida, it could have represented a path toward economic independence as well as solidifying further cultural and historical connections with various geologic flows significant to their culture. How much input the Haida had on the project, however, is a different matter.

As a corporate entity, the HSRC did not represent all Haida people. The

HSRC was a Vancouver-based company funded by the Old Massett town of the Haida, whose stated interest was in "developing progressive solutions to foster and restore ocean health."[82] Old Massett is only one village in the Haida Gwaii archipelago, and though it is one of the largest, its members do not speak for all Haida. The president of the Haida nation, Guujaaw, stated that the village was told the test would be of significant environmental benefit and bring back the salmon so crucial to their livelihood. "The village people voted to support what they were told was a "salmon enhancement project," Guujaaw later clarified, "and would not have agreed if they had been told of any potential negative effects or that it was in breach of an international convention."[83] After the test, however, the Haida Nation condemned the events, and the Council of Hereditary Chiefs and the Council of the Haida Nation stated they "had nothing to do with HSRC's activities."[84] Geographers Kate Gannon and Mike Hulme argue that Russ George, who is not a Haida Nation member, wrote large portions of the initial HSRC website, making "sense of the HSRC project through [his] own versions of Haida oral histories and stories, as well as through interpretations of Haida political ambitions and relationships."[85] George was intentionally trying to align Planktos "with the Haida name and proud identity of environmental stewardship," specifically by positioning iron fertilization as "an intentional act of nurture."[86] The initial HSRC website referenced the slogan "ancient wisdom + new science" and recontextualized a Haida story, "Salmon Boy—How the Haida and Salmon Are as One," interpreting it in a different fashion than in traditional narratives.[87] The HSRC was therefore a project conducted by non-Haida, off-island people and represented an aggressive act of cultural appropriation.[88] Planktos attempted to reduce the Haida to a strategic part of a "neoliberal narrative of restoring lost balance through for-profit, technocratic, unilaterally deployed management schemes."[89] Planktos, by way of the HSRC, used the Haida to further align their geoengineering project with Indigeneity, and through it a status deemed to be more "natural."

The Planktos incident is an example of how an ecomodern geoengineering company aligned itself with a First Nation to borrow their ethos and represent their goals as synonymous with Indigenous ecology so as to generate interest in unfounded claims about CO_2 sequestration. Planktos did so by claiming to have found a simple technological resolution to a complex environmental and social problem. In promoting iron fertilization and asserting that there is a profit to be made in selling carbon credits, they intentionally oversold the efficacy of their methods, which could only support increased phytoplankton growth, at best. At the same time, by trying to make the Haida into a metonym for what is natural, Planktos managed the representation of the Haida through the HSRC as part of a larger mobilization of "the idyllic chronotope to naturalize management ecology in the mode of green entrepreneurship, building a milieu that collapses human and nonhuman scales in its speculative annexing of the ocean."[90] Planktos tried to use the Haida as evidence for the "naturalness" of ocean fertilization, relying on a trope framing Indigenous people as "noble savages," closer to nature. They did so clearly for public relations purposes, even though their methods are derived from a lack of clear distinction between physis and nomos.

In the end, there is little evidence that the experiment succeeded. In 2013 the Pacific Salmon Commission reported a massive increase in the numbers of Pacific pink salmon, potentially influenced by the Haida experiment, possibly the result of natural algal blooms, or possibly due to other factors.[91] The Planktos experiment could have created conditions for the rise of pink salmon, but conditions in the north Pacific make it especially difficult to determine cause and effect. Marine scientists reported to the ETC group that the location "would be the perfect place to put iron to guarantee your patch would have more chlorophyll than the surrounding waters. In fact, it probably would be impossible to show that anything different from a natural bloom was caused by your iron delivery."[92] Despite claims that Planktos would share scientific evidence gathered from their research, the only study to be published at the time of writing is a single, loosely related conference proceeding. Because of the lack of transparency, the dearth of scientific basis for conducting the test, and the choice of location for performing the research, we are unlikely to know the results that could have been gleaned. Gannon and Hulme also doubt that "with no market or mechanism for verifying carbon credits from OF currently in existence, it remains unlikely that Old Massett's investment—let alone the 'guaranteed' profits from the project . . . will ever be returned to the village."[93] Perhaps most concerning about the Planktos experiment, however, is that it is representative of how geoengineering entrepreneurship could develop in the near future: with little oversight, with poor understanding of the foundational science, based upon a techne that can shape-shift to meet the needs of the designers, and rationalized by way of an ethical alignment with the world's most-marginalized communities. Naomi Klein also weighed in on the significance of the Planktos test, offering her opinion that "the results of Mr. George's test are clear: geoengineers proceed, caution be damned."[94]

The Planktos experiment allows us to see the possible future of geoengineering: for-profit endeavors, beyond the reach of international law, misinterpreting scientific research, and claiming a deep connection to natural forces, regardless of their ecomodern ambivalence for distinctions between physis and nomos. This "experiment" was an overly optimistic, exploitative version of science occurring beyond the scope of any institutional authority. Holly Jean Buck sympathetically portrays the Planktos event as "a *different type* of science [in which] the methods, funding sources, and intents were often a hybrid of conventional and unconventional, or professional and amateur, practices. The project was conceived by people who didn't have advanced training in science, but wanted to do the work."[95] While this enthusiastic rogue science experiment possibly hoped to build on $techne_1$ methods in search of new evidence, it is more likely that geoengineering was seen as an opportunity to prey on the historically and geologically disempowered as a means for investment. This pivot-point of confusion is often where geoengineering entrepreneurs focus their deliberative powers, and it is also where we see a caustic geologic rhetoric at its most evident. Geoengineering companies can warp or misrepresent scientific findings, muddy the differences between tried-and-true methods and experimental methodologies, and capitalize on the lack of clear cultural and theoretical demarcations between nature and culture. When working upon such ungrounded ground, more abstract techne are bound to flourish. Planktos is not the only company pursuing this geoengineering idea, and geoengineering appears to be ripe for capitalist speculation that employs the language of cybernetics and ecomodernism in the service of technologically specific solutions.

Geoengineering start-ups are likely to thrive in coming years, as while ocean fertilization companies may have temporarily reached the limits of their efforts, other geoengineering approaches are still in their developmental stages. Perhaps the best indication of this potential lies not in the large-scale projects funded by Bill Gates, Richard Branson, or Elon Musk but in the projects advocated by smaller, more nimble start-ups, each eager to advance their methodologies in search of capital backing or new investment. The seed money start-up accelerator Y Combinator, already behind such successful platforms such as DoorDash, Twitch, Instacart, Reddit, and Dropbox, announced a CDR-themed start-up project in 2018. They did so while proclaiming that "it's time to invest and avidly pursue a new wave of technological solutions to this problem—including those that are risky, unproven, even unlikely to work. It's time to take big swings at this. . . . These ideas press on the limit of what's possible, and we're not sure which side of that line they're on. . . . The frontier technology ideas presented in this RFS [request for startups] straddle the border between very difficult to science fiction. While these approaches are not our Plan A, we think it's time to get Plan B ready."[96] There might be no better description of $techne_2$ geoengineering approaches, pursued regardless of potential risk, advancing rogue science, and grounded by understanding of kairos that wants to "move fast and break things."[97] Perhaps most concerning of all, while Y Combinator has funded big, successful projects, like Airbnb and Coinbase, hundreds of less-successful start-ups receive large seed funds—sometimes with little oversight and sometimes failing to deliver on their promises. The risk of a failed tech start-up includes lost jobs and wasted capital, but a failed geoengineering start-up could mean environmental destruction and a bankrupt company that cannot be held liable. Further, the logics prevalent at most tech start-ups might uncomfortably echo the cybernetics-influenced, ecomodern belief that Comprehensive Designers can bring about a "good" Anthropocene. The following chapter delves into the legacies of cybernetics theory and how it has influenced both the articulation of Anthropocene and the development of geoengineering.

5

HOMO FABER

Six years before Paul Crutzen broke the taboo against discussing geoengineering within the scientific community, he—together with Eugene Stoermer—published the first paper to mention the concept of the Anthropocene. Crutzen and Stoermer's now well-known argument laid out some of the evidence for how humans have negatively and permanently impacted the earth: fossil fuel overuse, CO_2 emissions, population growth, industrial agriculture, nitrogen fertilizer misapplication, species extinction, extreme forest fires, deforestation, toxic pollution, ozone hole expansion, and more. They also noted that the Anthropocene was preceded by another, similar concept—the "noosphere," a term defined as "the world of thought," employed "to mark the growing role played by mankind's brainpower and technological talents in shaping its own future and environment."[1] Crutzen and Stoermer emphasized that a "real functioning noösphere" could prevent "major catastrophes like an enormous volcanic eruption, an unexpected epidemic, a large-scale nuclear war, an asteroid impact, a new ice age, or continued plundering of Earth's resources by partially still primitive technology."[2] Apparently, an operational noosphere is a necessary component to creating "a world-wide accepted strategy leading to sustainability of ecosystems against human induced stresses."[3] Strangely, the concept of the noosphere was not only mentioned in this one article but has repeatedly been referenced by other major Anthropocene commentators.[4] Even Rajendra Pachauri, the former chairman of the IPCC, has written about the "mental component of the earth system" and wondered whether "the mental sphere [can] evolve quickly and purposefully to a point where the future evolution of the system can be managed consciously toward a state of dynamic sustainability of the whole system?"[5] While few will recognize this reference, the concept of the noosphere has deep roots in vitalism, geology, and cybernetics theory.

These continual references to the noosphere connect with assumptions that position people not as homo sapiens but as homo faber, the toolmaking and tool-using species that has been responsible for the coming ecological catastrophe and is, perhaps paradoxically, the only hope to solve our ecological problems. These references also provide us with insight into the suppositions about humans being made in ecomodern Anthropocene theories: namely, they assume humans are somehow genetically predisposed to technological advancement and that therefore the Anthropocene can be blamed on a species and not certain members or certain cultures within that species. Here, a geologic rhetoric prompts us to investigate the deep-seated assumptions some Anthropocene thinkers have about humans' biological traits and how that species-thinking impacts geoengineering deliberation. Specifically, it points us to how references to the noosphere allow for the further rationalization (and naturalization) of geoengineering, as noosphere logic suggests that while humans have caused climate change, some exceptional members of the species—geoengineers—will be able to fix it. For Kathryn Yusoff, language like this tries to achieve an ideal "of the Anthropos as universal subject [where] the human needs to become both abstracted (from its previous forms of exclusionary humanism) and already populated in the form of the White Western master-subject."[6] This dangerous logic pervades geoengineering thinking and, comparable to the history of ecomodernism, has roots in early cybernetics theories. It also makes assumptions about homo sapiens' capacity for resilience. While rhetoricians have previously approached resilience from the framework of ecological thought, another form of resilience is gathering momentum, an engineering resilience focused on returning to a previous state as quickly as possible, thereby offering additional support to faster methods of geoengineering like SRM. We can see this resilience beginning to manifest in visual representations of climate change and in the comparative assessment of SPI. Offering caution about the potential for this logic to prevail, this chapter traces these emergent forces in geoengineering, beginning with the strange references to the noosphere in modern-day discussions of the Anthropocene.

SPECIES-THINKING IN THE ANTHROPOCENE

Like the term geoengineering, the term noosphere is a hybrid, a word comprised of the Greek for "mind" and the Latin for "globe." By some accounts, the idea of the noosphere is attributed to Eduard Suess, the Austrian geologist who also coined the term "biosphere." Suess does not appear to have use the word in print but rather hypothesized about the concept in *The Face of the Earth*, a 1904 book that offered up an early version of continental drift theory.[7] Direct references to the noosphere first occur in works by the philosopher and paleontologist Pierre Teilhard de Chardin, the French professor of mathematics Édouard Le Roy, and the Russian/Soviet geochemist Vladimir Vernadsky. Vernadsky often gets credit for popularizing the concept, but Teilhard was the first to explain it. For Teilhard, the noosphere is a mystical form of universal human consciousness, the sum of human thought, where technology is to be embraced in the inevitable journey toward the "Omega point," a final stage of a unified Christian enlightenment. Vernadsky's version is decidedly less mystical than Teilhard's, but more geologic, in that he positions the noosphere alongside the biosphere and the geosphere as the three major forces that structure and govern the earth. Vernadsky's works *Geochemistry* (1924) and *The Biosphere* (1926) were first published in Russian—limiting their global readership—and described how human thought impacts the world.[8] Vernadsky's noosphere was grounded in the assumption that not only were humans changing the earth, this process was "pre-formed by paleontological evolution."[9] In his account, humans are a dominant geologic force beholden to the inevitable evolutionary processes that structure life. As Vernadsky explains it, the noosphere could act as a mechanism for understanding and guiding that power.

Framed as an evolutionary process that controls human beings, the noosphere implies that humans are inexorably bound to continue advancing technology throughout time. Technological development is not just desirable but, in some sense, encoded within our evolution. Much like how beavers are hard-wired to make dams, or birds are biologically determined to make nests, humans are coded to develop increasingly complex technologies—with the goal of constant human advancement. Commenting on Vernadsky's writing, Bertrand Guillaume explains that Vernadsky was "convinced of the inescapable potential of science and knowledge to point civilization in the direction of universal progress."[10] For Vernadsky, humans represented a manifestation of cognitive development so intense that the earth's future must naturally be powerfully interconnected to that growth; the technologies we surround ourselves with are extensions of our cognitive capacity and comprise the noosphere. Clive Hamilton and Jacques Grinevald explain that "both Vernadsky and Teilhard understood evolution as a process directed towards ever-higher levels of cephalization (the development of the central nervous system culminating in the continuing growth of the human brain) and consciousness. They saw the development of intelligent life, including technology and scientific research, as a law inscribed in a cosmogenesis (Teilhard) or in the evolution of the planet Earth's biosphere (Vernadsky)."[11] By linking human technological development to cognitive evolution, references to the noosphere position the Anthropocene as an inevitable stage of biological development where human technological progression is naturalized as a genetic predisposition and geologic force. Mentioning the noosphere also does two main things for Anthropocene arguments. First, these references work to naturalize all previous human activity into an undifferentiated, genetically predetermined mass. The noosphere bolsters the claim that any human action resulting in modern-day environmental damage was inevitable; it also fails to appoint blame to colonial, Western, industrial modernity for that destruction. Secondly, the logic of the noosphere provides a rationale for what deliberative decisions humanity should take in the future to solve the current climate crisis. This logic suggests that humans should accept their cognitive and technological development as biologically fated and likewise embrace any technologies that help us control our genetic and geologic limitations. Repeated incantations to the noosphere are not coincidental, as the idea has been closely connected to both cybernetics and ecomodernism—worldviews that are often used to support the Anthropocene and rationalize geoengineering.

Frequent references to the noosphere suggest that Anthropocene thinkers believe environmental harm is the result of the genetic predisposition of all homo sapiens, in part because the noosphere describes the capacity of human thought—capacity that exists at a species level. This framework ignores the complex nuances of *actual history* and positions humans as an undifferentiated biological species. Defined in this way, humans are metamorphosed into the causal effects they have had upon the earth; furthermore, when viewed through a lens of deep history, where humans are represented by the qualities that make them distinctive from other animals, such species-thinking appears as an apolitical, genetically oriented rationalization. For political theorist Matthew Lepori, these framings of the Anthropocene "narrate the rise of the human species, from being merely *erectus* and relatively benign to *sapiens* and all-too-powerful. Scientists demonstrate that with each successive leap in resource exploitation and technology, beginning 8,000 years ago with agriculture and intensifying greatly in the nineteenth century with fossil fuels, industrialization, and synthetic nitrates, human intervention into nature dramatically increased in scale."[12] Variations of "species-thinking" have been demonstrated in a number of Anthropocene theories, though none has received as much attention as Dipesh Chakrabarty's 2009 "The Climate of History: Four Theses."[13] Here, Chakrabarty gives human genetics a major role in historical narrative, thereby minimizing the intentional decisions that have occurred to create climate change. His argument erases the widespread and pervasive disempowerment of much of the world's population by flattening the causal forces of humanity and ignoring historic and structural inequalities. Most of humanity has not been party to much of the industrial fossil economy, and "hundreds of millions [still] rely on charcoal, firewood or organic waste such as dung for all domestic purposes."[14] Many humans have had (and still have) a carbon footprint of almost zero and "from the consumption perspective, globally, the 20 per cent of the population with the highest consumption levels is likely to account for more than 80 per cent of all human-induced GHG emissions and an even higher proportion of historical contributions."[15] With such variance within human responsibility for climate change, the move to lump everyone under the banner of one species becomes an intentionally depoliticizing one that seeks to contravene current and historic social and economic inequalities.

References to the noosphere are ultimately made to support biological determinist arguments about human capacity. *Of course* we would negatively impact the sustainability of our natural world, as that outcome is just a result of humans' cognitive aptitude and our potential to change our environment. Our capacity to make world-altering technologies, for better or worse, is simply who we are. Christophe Bonneuil defines this as framework as *the dominant narrative about the Anthropocene*, one that describes how our species has evolved into a major geologic force starting a few million years ago when early hominins discovered tools, made conscious use of fire, and achieved greater cognitive power. The narrative frames human history "as the ambivalent odyssey of Man from hunter-gatherer to telluric force, as the epic confrontation between the 'human species' and the 'Earth system.'"[16] Invocations of the noosphere provide a species-level explanation as to why climate change has occurred, thus absolving any particular culture from its rightful responsibility, but they also provide for a second, equally distressing rationale. In projecting the noosphere into the future, Anthropocene commentators—some of whom are also geoengineering researchers—posit that only certain forms of research and deliberative planning ought to be considered in the orchestration of a "good Anthropocene." This ecomodernist narrative of the future fits neatly alongside Vernadsky's noosphere, as both situate humans as able to overcome the environmental problems they have also caused. We should not feel bad about being able to change the environment, this argument contends, not only because humans have always done so but also because that path is the only option we have left.

Emphasizing the existence of the noosphere also further naturalizes geoengineering as a deliberative option. *Of course* we will create positive, intentional changes that result in a managed, controlled, and ultimately better environment. Humans could not have done anything but create the conditions for technological destruction, because it is a precondition of our species, yet humans can create the conditions for redemption because we are good at technological innovation. With the right deliberative strategy, humans will create a better future—a "good Anthropocene." Bonneuil explains this as a trajectory showing "man moving from environmental obliviousness to environmental consciousness, of Man equaling Nature's power, of Man repairing Nature."[17] Here, geoengineering a good Anthropocene becomes the naturalized outgrowth of human biological and cognitive resilience and suggests a particular kind of capacity that prioritizes ecomodernist geoengineering research strategies. The ecomodernist movement has not been shy in employing the noosphere in rationalizing its environmental positions. Breakthrough Institute–affiliated journalist Andrew Revkin has made many of these connections apparent. Revkin contends that scientists agree that Vernadsky was right and that "global society, guided by science, would soften the human environmental impact, and earth would become a 'noösphere,' a planet of the mind," where life is ruled by reason.[18] These words echo those in Brand's *Whole Earth Catalog*, the "Ecomodernist Manifesto," and Vernadsky's own claims that if humans were to understand that their strength is derived from collective intelligence, "an immense future [would be] open before him in the geological history of the biosphere. . . . Therefore we may face the future with confidence. It is in our hands."[19] Yet, there is a troubling discrimination at play in these deliberative arguments.

At the core of these appeals to the noosphere is a positivist, modernist faith in the ability of rational science to innovate our way out of impending climate change problems. When applied to the past, appeals to the noosphere are used to blame the climate crisis on homo sapiens' inherent traits and offer an evolutionary argument that depoliticizes and ignores difference. When applied to the future, however, this same rationale suggests that we have only one way out of our species' inevitable descent toward climate destruction. Only by placing control of the earth in the hands of the best and brightest geoengineers will humans be able to counteract the genetically hard-wired predilections of our species toward waste, greed, and environmental destruction. Crutzen and Stoermer make this argument clear in their paper introducing the Anthropocene, suggesting that "an exciting, but also difficult and daunting task lies ahead of the global research and engineering community to guide mankind towards global, sustainable, environmental management."[20] That management, in short, is likely to result in geoengineering, and only certain kinds of human thought (higher forms of "cephalization") are going to be invited to create this management. Geoengineers, we are led to believe, will succeed where others have failed.

HOMO FABER

Along with appeals to the noosphere, the earliest articles defining the Anthropocene focused on how scientists' and engineers' efforts could shape the future for the better. Crutzen considered the intimidating task that "lies ahead for scientists and engineers to guide society towards environmentally sustainable management during the era of the Anthropocene. This will require appropriate human behaviour at all scales, and may well involve internationally accepted, large-scale geo-engineering projects, for instance to 'optimize' climate."[21] His interpretation sets scientists and engineers apart from the undifferentiated population of the world, a mass of people who are named as causing the problem. Crutzen and other Anthropocene thinkers have staged Western science as detached from the supposedly "biological" conditions that have generated the Anthropocene, removed from the inevitable forms of human action that have led to the present moment of environmental destruction. In their formulation, science is a "deus ex machina that was not part of the cultural–political–economical nexus that made the Anthropocene, but which will now guide humankind and save the planet."[22] Scientists are figured as rational leaders of irrational people whose profligate use of natural resources has caused climate change. This duplicitous construction depicts a species-level distinction that has come to dominate Anthropocene thinking and geoengineering planning: while homo sapiens are the cause of the world's climate problems, a special force of geoengineer homo sapiens is the solution. Alvin Weinberg would no doubt be proud.

But to further naturalize geoengineering, humans needed to be rhetorically reconstructed as biologically predetermined to solve social and political problems through advanced technology. The definition of homo sapiens frequently referenced in Anthropocene commentary, and indirectly in work on geoengineering, is that of homo faber, or the tool-using and toolmaking human. Homo faber is an ancient concept first committed to writing by Appius Claudius Caecus, who wrote, "Homo faber suae quisque fortunae," which roughly translates as "Every human is the maker of his or her destiny."[23] The phrase has also been used by several theorists throughout the twentieth century, from Hannah Arendt to Kenneth Burke, but also by Henri Bergson in *Creative Evolution*, a text that heavily influenced Vernadsky and Teilhard. For Vernadsky, "a new geochemical fact of capital importance is manifest. . . . Man has introduced a new form of action of living matter on the exchange of atoms of living matter with inert matter. These are no longer the only necessary elements of production and the formation of living matter which enter into play here and change their molecular structure. These are necessary elements to the techniques and creation of civilized forms of life.

Here, man acts not as *Homo sapiens*, but as *Homo faber*."[24] And for Bergson, renaming homo sapiens as homo faber emphasizes human beings' defining trait—the ability to make and use tools. He argued that "if, to define our species, we kept strictly to what the historic and the prehistoric periods show us to be the constant characteristic of man and of intelligence, we should say not Homo sapiens, but *Homo faber*. In short, *intelligence, considered in what seems to be its original feature, is the faculty of manufacturing artificial objects, especially tools to make tools, and of indefinitely varying the manufacture*."[25] In Vernadsky's and Bergson's arguments, humans are biologically predisposed to change the conditions of the world through techne and technology. Our tool use may be an imperfect science, Bergson claimed, but it allows us to deliberate about the future and asks us to consciously choose between multiple designs, options, and effects.

We can easily comprehend how the idea of homo faber might be attractive to those looking to rationalize geoengineering as the means to solve the problem of climate change. Identifying humans in this way emphasizes how our capacity for tool making and tool use can be blamed for the abuse of fossil fuels, further naturalizing climate change as inevitable and resultant from our innate abilities, while also removing liability from that same relationship—a clear continuation of the noosphere ideation. Naturalizing the human capacity to create tools likewise establishes our inclination to develop additional technologies to fix the problem of climate change through geoengineering. The concept of homo faber allows these narratives to function together, where human genetic capacity is both the cause of *and potential solution to* the increased CO_2 emissions and therefore the climactic instability of the Anthropocene.

The framing of humanity as homo faber also splits humanity into factions that are continually emphasized in geoengineering and ecomodernist discourse. On the one hand, humans are portrayed as "competitive, selfish and predatory," attributes that fit well with the belief that we will always self-interestedly make use of the earth.[26] This is the homo faber of resource exploitation, excessive capitalism, and colonial power—a narrative that Daniel Chernilo believes connects with the "idea of the anthropos as a selfish being whose only drive is the constant satisfaction of its multifarious needs through an incessantly instrumental attitude towards everything and everyone that surrounds it."[27] Christophe Bonneuil adds, "This narrative pictures society as ignorant, passive and stuck in 'cognitive dissonance'" because people are seemingly unable to change their destructive patterns of behavior.[28] This perspective assumes humans have the capacity to make large-scale infrastructure or lifestyle decisions and implement them en masse, but continually fail to do so. Climate change is thus framed as a social problem caused by humans who act without reason. On the other hand, humans are also believed to be "evaluative, cooperative and reflexive beings,"[29] and so we see a second meaning of homo faber emerge that believes people are able to, in Paul Crutzen's words, "guide society towards environmentally sustainable management during the era of the Anthropocene."[30] Clive Hamilton argues that the ecomodern version of homo faber frames human beings not as "destructive creatures—and certainly not sinful ones as some greens imply—but creative, ingenious and basically well-meaning. If it is our destiny to inherit Earth, then the arrival of the Anthropocene is the fulfilment of that destiny."[31] These human traits are also highlighted in most descriptions of a "good Anthropocene," and when applied to the biological predispositions of homo sapiens, they come freighted with numerous political, technological, and cultural implications. This belief in the two natures of humanity is anything but new, but its application to justify geoengineering is now becoming evident.

An excellent example of the dual construction of homo faber shows up in ecomodern writer Mark Lynas's book *The God Species*, wherein Lynas articulates homo sapiens' current godlike ability to control the fate of the planet.[32] He praises Stewart Brand's prescient thinking in the *Whole Earth Catalog* and argues that "playing God (in the sense of being intelligent designers) at a planetary level is essential if creation is not to be irreparably damaged or even destroyed by humans unwittingly deploying our newfound powers in disastrous ways."[33] Lynas hails the belief that in order to combat environmental degradation, those with the foresight to envision the future should take on the role of planetary designers, a position strikingly similar to Buckminster Fuller's Comprehensive Designer. Lynas is profoundly frustrated that humans are oblivious of how they are already changing the planet and is astounded at "how blissfully unaware of this colossal transformation we remain. We are phenomenally, stupendously, ignorant. As if God were blind, deaf, and dumb, we blunder on without any apparent understanding of either our power or our potential."[34] In order to combat global warming, Lynas sees geoengineering as a natural outgrowth of human capacity and one of the few major options that we have to control the worst repercussions of that same capacity. This kind of discourse ignores the actual histories of political and economic inequality that have led to climate change, assumes a deeply Western, rationalist bias, and focuses heavily on idealistic solutions rather than apportioning blame through appropriate, justice-oriented measures. That kind of critical self-reflection, ecomoderns would presumably argue, will not help us fix our present situation.

Articulations of homo faber have become more and more pronounced in recent ecomodern descriptions of humanity, in part because focusing on the many ways in which we are "competitive, selfish and predatory" does not provide opportunities for confident, deliberative decision making. Ecomodern thinkers needed to rhetorically reinvent homo sapiens as a species whose innate inventive impulses will lead toward productive environmental rehabilitation despite its worst tendencies. A pessimistic, apocalyptic narrative of the Anthropocene places limits on human potential, but incorporating an ecomodern narrative of a "good Anthropocene" encourages human technological creativity to solve humanity's environmental ills. This belief is not often explicitly stated, but its appeals to the noosphere and references to homo faber have helped to establish both of these dimensions as cornerstone features of deliberative geoengineering rhetoric. The solutions that can be envisioned under these concepts, however, are guided only by a specific kind of technological development. Here, rational science is the force that will save the planet, and it will be fostered through collective human intelligence and the unique capacities of human resilience. The forms of resilience that could dominate within this model, however, are likely to buttress faster, cheaper, and more risky geoengineering solutions.

CAPACITIES FOR RESILIENCE

Because we are now actively contending with the effects of climate change, it is resilience, and not prevention, that has become the new guiding focus. The idea of resilience has been used by commentators to describe how humans can sustain their lives and livelihoods while dealing with ongoing climate change. The IPCC defines resilience as "as the ability of a social, ecological, or socio-ecological system and its components to anticipate, reduce, accommodate, or recover from the effects of a hazardous event or trend in a timely and efficient manner."[35] Resilience to the effects of climate change is thought to be necessary, and the development of rooftop solar panels, fire-resistant landscaping materials, gray water systems, and alternative transportation solutions are all resilience-oriented strategies that can be implemented—and there are numerous resilience plans being drafted or put into effect, primarily at the level of individual industries or municipalities. However, when considered at a global level and the scale of geoengineering, the concept of resilience includes assumptions about humans not unlike the ecomodern Anthropocene commentary. For example, any notion of resilience must necessarily assume certain capacities on the part of the group being affected by an undesirable change. Just as an ecologist might study how an owl population has been impacted by the lack of old growth forest and therefore makes suppositions about the capacities of those owls to adapt, geoengineering advocates make hypotheses about the capacities of humans to be resilient in the face of a changing climate. Of course, human capacities for resilience are different, as we have surrounded ourselves with the products of our invention as means to increase our ability to withstand negative change. Our resilience, therefore, is of a different kind. Any argument for the role geoengineering could play in human resilience necessarily makes hypotheses about human capacities, both those that hinder our ability to react to climate change and those that allow us to create and deploy the tools of our invention, geoengineering included.

In the context of global warming, human capacity is often represented as limited by other biological traits that hinder our ability to deal with (or fully understand) climate change. That is, we can see how human capacity, understood as a set of psychological limitations, defines how we have been responding to climate change. Albert Lin has argued how the human capacity to react to climate change has been restricted because of the psychological heuristics, emotional influences, and cultural cognition perspectives that we hold. Lin notes four prominent cognitive shortcuts—optimism bias, hyperbolic discounting, outrage, and the availability heuristic—that can reinforce how people subconsciously convince themselves of the significance or insignificance of climate-related risks.[36] Typically, these are seen as species-related hindrances to our resilience. Daniel Kahneman, an influential economic psychologist perhaps best known for the book *Thinking, Fast and Slow*, believes that these cognitive biases are unlikely to be overcome anytime soon.[37] Kahneman has said that he is not optimistic of our ability to deal with climate change because "no amount of psychological awareness will overcome people's reluctance to lower their standard of living."[38] Of course, and as many critics have argued, people are more than their generalized psychological traits. The technological environments that we have created and surrounded ourselves with also play a role in our capacity to be resilient against large-scale existential threats. Any model that assumes humans have the ability to create symbols and technologies must necessarily take those distributed capacities into consideration.

For rhetoricians Nathan Stormer and Bridie McGreavy, capacity is a preferential way to articulate the human potential for action, because it considers the smart environments we surround ourselves with. They prefer the idea of capacity over that of agency, where agency is often defined as the ability to act in typically human-to-human relations. Here, agency might be more directly connected with our psychological hindrances. Stormer and McGreavy define rhetorical agency as "the instrumentality of force" and protest that the concept isolates agency as something determinable and locatable within a body or institution—a notion they argue is an inaccurate way of theorizing the potential and ability to act.[39] In order to identify a more ecological and relational version of agency, they promote the idea of "rhetorical capacity," where capacity "defines the limits of what can be done," according to the material and ecological affordances provided in a situation.[40] Their goal in shifting focus to capacity is to emphasize "the ecological entanglements between entities over the abilities that are inherent to humans."[41] Understanding the impact of this shift for geoengineering is tricky, however, as "separating capacity from agency . . . involves scrutinizing the assumption that human limits define rhetorical limits," something that has been a challenge for rhetoricians to do.[42] To perform this scrutiny, Stormer and McGreavy look back to Kenneth Burke's definition of the human. Namely, Burke describes man in *Language as Symbolic Action* as "the symbol-using (symbol-making, symbol-misusing) animal, inventor of the negative (or moralized by the negative), separated from his natural condition by instruments of his own making, goaded by the spirit of hierarchy (or moved by the sense of order), and rotten with perfection."[43] Too often, they argue, rhetoricians use Burke's definition to limit rhetoric to the human sphere, because the field has traditionally seen "the mark of humanity as its organizing interest and yet also the means by which humans transcend and remake the world and themselves."[44] Considering rhetorical capacity opens up this species-based limitation in ways that Burke, and others, could not have anticipated.

Stormer and McGreavy's articulation of capacity as a relational potential supports their definition of rhetorical resilience, where resilience describes a species' capacity to be "responsive, adaptive, and persistent" to changing situations and environments.[45] They intentionally draw this definition from the field of ecological thought and ecological systems theory, where resilience generally describes adaptive abilities in changing and uncertain environments.[46] Stormer and McGreavy argue that a being's surrounding material conditions establish the realms of capacity for which rhetoric can be generated and that "different materialities set the field of potential and condition diverse rhetorics' emergence from the broader environment."[47] They point to other species' ecologies, noting that, for example, a "shellfish bed's resilience is its ability to absorb disturbance and still act like, feel like, and look like a clamflat or mussel bed."[48] This definition of human resilience describes it as that capacity that allows for acclimatization and responds to the complex networks of which it is a part; it is a distributed sense of capacity and therefore a networked resilience. This definition also sits alongside that offered by Elizabeth Flynn, Patricia Sotirin, and Ann Brady in *Feminist Rhetorical Resilience*, where resilience is a "a process of rhetorically engaging with material circumstances and situational exigencies."[49] Like the definition of kairos offered earlier, this conception of resilience entails an ongoing responsiveness to material and social relations with the goal of persevering through adversity.

Unfortunately, while the examples that Stormer and McGreavy give are useful in understanding an idealistic rhetorical resilience, the dominant species-thinking perspective in line with most ecomodern geoengineering commentary would argue that humans exist with a different order of capacity within the ecologies they inhabit—very much unlike a clam in a clamflat. As suggested by the idea of the Anthropocene, humans have changed their ecosystems, and the capacity to make those changes is likewise assumed to be part of what makes our species unique. Homo faber invents tools and then uses them, and in that use, changes its environment; as Mark Lynas suggests earlier in this chapter, homo faber is a form of "God species." Ecologists have recently created a name for the kind of position that humans play in their ecosystems, one that reflects these same Anthropocene assertions—a hyperkeystone species. A keystone species has a large impact upon its ecosystem, especially considering its relative abundance. Robert Paine first defined the concept in 1969 and explained how a keystone species, like large, predatory mammals, can assert an outsized influence where it lives.[50] An ecosystem relies upon several keystone species to sustain other ongoing relationships. In contrast, a hyperkeystone species manipulates and exploits other keystone species, altering their relationships and ecologies. Hyperkeystone species can modify the most fundamental conditions of other species and, perhaps more significantly, intentionally and unintentionally change the ecological conditions for themselves in the process.

Ecological systems theory approaches the power inherent in a hyperkeystone species with a plea to researchers to study humans and nature "as an integrated whole, not as separate parts."[51] In "Social-Ecological Resilience and Biosphere-based Sustainability Science," Folke et al. argue that "the social-ecological resilience approach emphasizes that humans and well-being fundamentally rest on the capacity of the biosphere to sustain us, irrespective of whether or not people recognize this dependence. It also makes clear that it is how we choose to view our relation to the biosphere in the globalized world and how we choose to act that will guide people in diverse contexts and humanity as a whole to sustainability or not. . . . For the sake of humankind, it is time to start accounting for and investing in natural capital, in active stewardship of landscapes and seascapes, and in the new global context of the Anthropocene."[52] This eloquent explanation is as much an ambitious appeal as it is an articulation of how humans are connected to the ecologies we inhabit. Unfortunately, while ecological systems theory is meaningful for those who believe it to offer an ethic for the Anthropocene, it may not offer enough explanatory power for those who see homo faber as offering the most accurate depiction of human capacity. It also may not guide the development of a networked deliberative response to intensifying climate change. Even those who espouse ecological systems theory can easily advocate for geoengineering a "good" Anthropocene. Case in point: Folke et al. frequently cite Erle Ellis, a specialist in biogeography and one of the coauthors of "An Ecomodernist Manifesto." Ellis has advocated for geoengineering in multiple locations. In *Anthropocene: A Very Short Introduction*, Ellis warns, "the less that is done now, the more societies will be willing to do for a cooler climate in the future. And geoengineering may turn out to be the surest way to achieve this."[53] But, Ellis also explains, "our unprecedented and growing powers also allow us the opportunity to create a planet that is better for both its human and nonhuman inhabitants. It is an opportunity that we should embrace."[54]

Conceptions of resilience from ecological systems theory may not be able to account for arguments about the power of a hyperkeystone species like homo faber, or the version of resilience it brings. The question of a harmonizing resilience could be replaced with concerns about how a single dominant species defines, manufactures, and performs its own resilience. Folke et al. suggest that an idealized version of ecological systems theory could approach the Anthropocene with "a process engaging people to collaborate across levels and scales and with shared visions and creativity framed by proper institutions, continuously learning and gaining experience and building capacity to live with change, adapt, and transform."[55] This framework aligns with Stormer and McGreevy's argument for an ethical form of networked, ecological rhetorics, in that it takes an inclusive, historical, and collective perspective on resilience. However, our future capacity for resilience could also be formed out of less-comprehensive, simplified definitions of resilience—one built upon a notion of ecological capacity that emphasizes homo faber's technological prowess.

How might a tool-using hyperkeystone species respond to negative changes in its ecosystem to make itself more resilient? One solution could be to make decisions guided by a notion of resilience defined as the ability to withstand external changes to existing and ongoing relationships. For climate change, we might cease all activities that result in the release of GHGs and create negative carbon industries, because such a decision would sustain biodiverse relationships and productive nutrient cycling in a global ecology. That change would be ideal and would follow principles in ecological systems theory; however, such a change would also require sweeping alterations in all sectors of industry and economics over the coming decade. In contrast to this version of resilience, a hyperkeystone species could adopt more time-sensitive and efficient resilience, one also defined by ecological science. The dominant definition of resilience in ecological science comes from the work of Crawford Stanley Holling. Holling defined resilience as "the persistence of relationships within a system," a definition based upon a system's relational capacity to sustain an existing network.[56] However, Holling noted that this measurement could be conducted in two different ways: through "ecological resilience" or "engineering resilience."[57] Ecological resilience "is a measure of the ability of these systems to absorb changes of state variables, driving variables, and parameters, and still persist."[58] The measure of ecological resilience focuses on the "magnitude of disturbance that can be absorbed before the system changes its structure by changing the variables and processes that control behavior."[59] This version of resilience closely aligns with Stormer and McGreavy's work because it takes into account numerous relational variables that can be defined by different attributes.

Conversely, engineering resilience "concentrates on stability near an equilibrium steady state, where resistance to disturbance and speed of return to the equilibrium" are the variables used to measure resilience.[60] Engineering resilience is often measured by fewer factors and instead emphasizes the time it takes a system to reassert a previous condition. Arguments for SRM geoengineering tend to rely on an engineering definition of resilience, because along with a kairos of delay, a lack of stasis, a subject matter expert articulation of cost-benefit forms of risk, and a tendency toward discipline-specific research, it appears unlikely that humans will be able to agree upon (and take) any significant action on climate change until a time when the effects of global warming have become undeniable, aggressive, and systemic. The passing of numerous climate tipping points seems to suggest the same. An engineering version of resilience also might question why any adaptation is necessary, especially since adapting to climate change will bring vast economic hardship, humanitarian crises, and require wholesale changes of lifestyle and energy production for the Western world. As a hyperkeystone species with the capacity to change the conditions of its own ecosystem, homo faber could alter one of the most fundamental drivers of ecosystem disturbance—climactic temperature—through a technological intervention. This intervention could be guided by the pressure to act before widespread adaptation is necessary and driven by a utilitarian engineering interpretation of kairos. Such a suggestion fits with representations of humans as a toolmaking and tool-using hyperkeystone species, endowed with the capacity to change the circumstances of its existence through technological development. Geoengineering forms of resilience are likely to be driven by these simultaneously cynical and idealistic beliefs in human capacity—the notion that we will do little about climate change until, forced to either accommodate or respond, we respond primarily out of our technological capacity for resilience. This response is being led by a small geoclique, largely situated in wealthy Western countries, who have rationalized their scientific paternalism through claims about the noosphere—a quasi-mystical cybernetics-influenced concept that contains numerous assumptions about the predetermined capacities of human beings. While this trend is certainly disturbing, it does not exist only within the occluded references made in scientific journals. This form of resilience can also be found in the most dominant visual representations of climate change.

VISUALIZING GEOENGINEERING RESILIENCE

Engineering resilience could become the dominant model for future responses to climate change, building from the persuasive deliberative forces already promoting SPI. Ironically, this form of resilience is being tacitly supported through an emphasis on GHGs as driver of average global temperature—and by visuals depicting the same. These visuals typically show GHG emissions plotted against time, and the prevalence of this relationship has turned them into a causal enthymeme, one that further marks homo sapiens as a unique species with the power to change the planetary ecosystem. Those who present this causality through science communication efforts hope that its directness will spur people to take increased responsibility for that power through appropriate collective action. Ideally, this action would push humanity toward full carbon neutrality, though admittedly it would involve, as the IPCC explains, broad transformations in "industry; transport; buildings; and agriculture, forestry and other land-use (AFOLU) sectors," transformations that appear unlikely at present.[61] Without such widespread changes, the more probable form of resilience that ecomodernists believe could occur is a geoengineering solution created by a hyperkeystone species. That is, we could see a relationship emerge where more information about climate change confoundingly generates more momentum for geoengineering—an outcome that aligns with arguments about geoengineering as a moral hazard that are discussed in the next chapter.

The most well-known visual depiction of global warming is the Keeling Curve, a simple graph of CO_2 levels over time based on observations of CO_2 at the Mauna Loa Observatory, beginning in the International Geophysical Year of 1957–1958. Figure 1 depicts a simplified representation of the Keeling Curve, using ice-core data from before 1958 and showing Mauna Loa data from that point onwards. I have removed specific dates and measurements from this visual, as I want to highlight the larger trend that shows GHG emissions rising over time, where the rate of increase is also growing. All the visuals shown here will be similarly stripped of detail, in order to emphasize the larger trends concerning resilience at play. The Keeling Curve has so far provided startling evidence for human-caused greenhouse gas escalation and has been called by Naomi Oreskes, noted science historian, "one of the most important pieces of scientific work of the twentieth century."[62]

The Keeling Curve visual has been reproduced to such an extent that it has now become a metonym for understanding the relationship between the escalation of human action, CO_2 emissions, and the dramatic change to the climate. Often, it has been plotted against temperature increase to show the causal relationship between the two. Increases of CO_2 over the last hundred-odd years suggest a shock to Holocene stability, a claim supported by the graph's recent verticality. Other visuals use this knowledge as a starting point and pick up on the effects from this shock. A good example is an image created by Robert Rohde, based on data from the IPCC's *Special Report: Global Warming of 1.5°C* (figure 2). Rohde is a scientist with Berkeley Earth, an independent nonprofit group of scientists engaged in rigorous analysis of climate data. Rohde's graph uses CO_2 emissions/yr., represented in $GtCO_2$, instead of CO_2 ppm, and the graph highlights the vast difference between a "business-as-usual" emissions path, where humanity fails to slow CO_2 emissions, and other, positive outcomes that restrain global warming to "just" 1.5°C. The distinction between current trends and the multiple alternate futures where we avoid the worst effects of climate change is hard to miss. This chart, like others, stresses the importance of neutralizing CO_2 emissions by

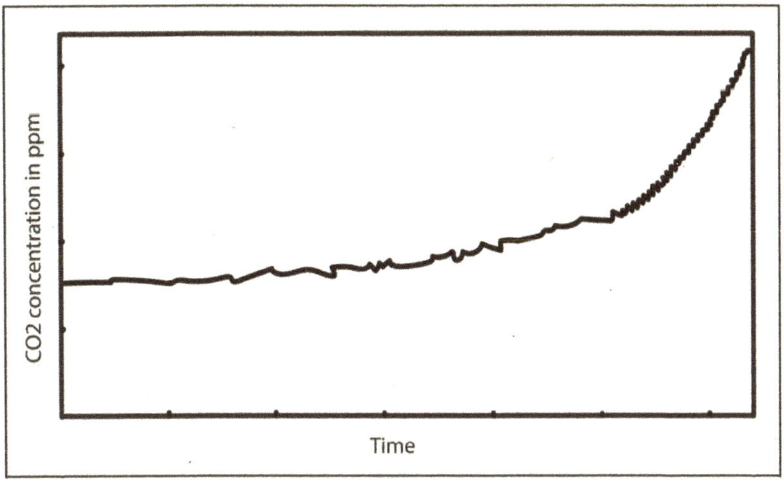

Figure 1. A simplified version of the Keeling Curve, representing 413.76 ppm, taken from Scripps Institute of Oceanography.

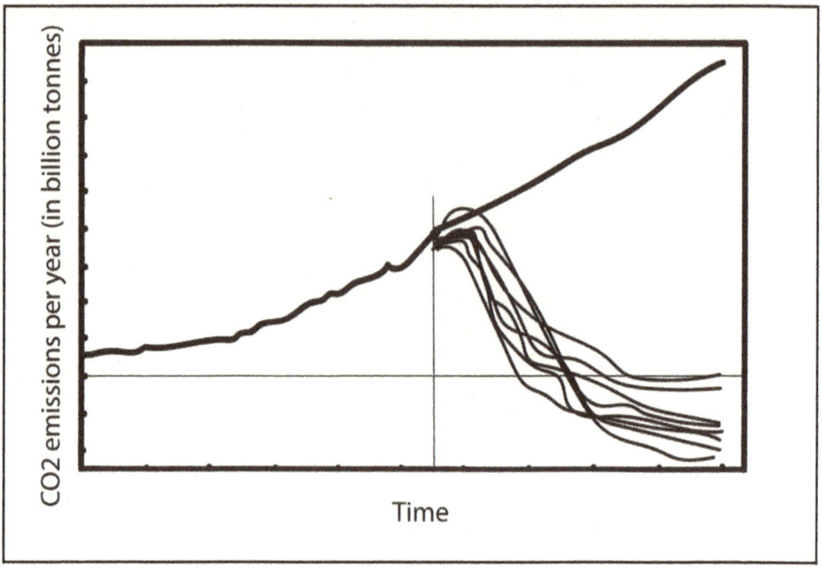

Figure 2. A simplified version of Robert Rohde's "Carbon Dioxide Emission Scenarios for 1.5°C of Warming." The thin vertical line represents the present day; the strong line to the left of the present-day shows historic CO_2 emissions. The strong line to the right signals business-as-usual emissions. Possible future scenarios for achieving only 1.5°C are represented by the multiple, branching lines to the right of the present.

around 2050 and then creating pathways toward negative global emissions. The lines moving forward from the present (the vertical line) show the kind of radically reduced CO_2 emissions that will keep global warming to within 1.5°C. In short, this is a visual representation of paths toward resiliency, as we strive to return to a state of preindustrial atmospheric CO_2.

In order to achieve one of the future pathways that IPCC recommends, the global output of CO_2 would need to reach peak emissions in relatively short order. This maximum release of CO_2 emissions needs to happen soon, and then decrease each year until we achieve global negative carbon output. "Barring unforeseen and transformative technological advancement," Walsh et al. have affirmed, "anthropogenic emissions need to peak within the next 10 years, to maintain realistic pathways to meeting the COP21 (aka, the 2015 United Nations Climate Change Conference) emissions and warming targets."[63] I take "transformative technological advancement" to mean either a miraculous energy source or sophisticated and precise CDR or SRM geoengineering.

For these representations, time is a key determinant in understanding the amount of CO_2 that can be released before global warming becomes devastating to planetary ecosystems. Each of these visuals stresses the uniqueness of our current situation: the dramatic rise in GHGs, the shock to our global ecosystem, and the necessity of responding to that increase in short order. Like engineering versions of resilience, these visuals emphasize the lack of time we have to return to climate equilibrium, and like rationales for geoengineering, they stress the difficulty of non-geoengineered futures to accomplish this task. Another visual provided by the communication outlet for the European Climate Foundation, *Carbonbrief*, makes this clear (figure 3). Here, the most important factor in avoiding the worst of human-caused climate change is the time it takes to reduce global CO_2. Again, most researchers agree that CO_2 emissions need to be reduced dramatically by about 2050, with full carbon neutrality occurring thereafter, followed by atmospheric GHG reductions. This timeframe for action is how most understandings of resilience articulate the parameters for what could/should happen and is becoming an agreed-upon, but often unspoken, response to the shock caused to our global ecology by industrial modernity. Resilience is here measured as the time it will take to return our earth system to "normal."

To achieve a radical carbon emission drawdown, we would need one of those "unforeseen technological advancements" that Walsh et al. suggest, because current pathways for emissions will simply send the planet sailing into

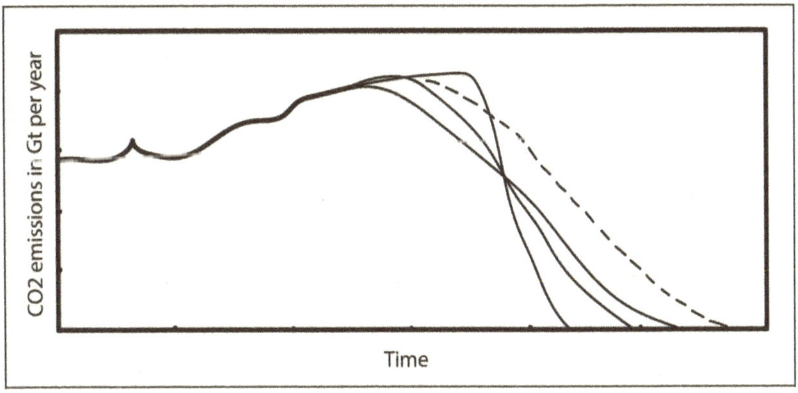

Figure 3. A simplified version of the "Carbon Crunch" diagram from Zeke Hausfather's description of when carbon emissions could peak in order to hold global warming to only 1.5–2.0°C. The point where one line splits into many shows possible paths forward for CO_2 drawdown in order to ensure that the planet will not warm by more than 1.5–2.0°C. The dashed line represents an 800 Gt carbon budget, peaking in 2020.

1.5°C or worse. What resilience to this coming temperature increase might look like from a geoengineering perspective is represented by figures 4 and 5. Figure 4 shows Chen and Tavoni's assessment of how direct air CO_2 capture (DAC) could contribute to a "base case" or "optimal mitigation pathway"—a scenario that could keep the earth from experiencing more than 1.5°C change in global temperature. DAC is one of the less risky forms of geoengineering, one focused on removing the CO_2 from the atmosphere to bring ppm down to preindustrial levels. It is also a form of geoengineering most in line with ecological resilience. Unfortunately, if numerous other features of a carbon drawdown are not also established, even a huge investment in DAC "would result in an *increase of carbon concentrations* at the end of the century of 540 ppm."[64] In general, DAC expresses a weak engineering resilience. An alternate geoengineering representation is offered by Robock et al. (figure 5). They modeled stratospheric aerosol injection against temperature (since SPI will not affect atmospheric CO_2). Here, SPI expresses a very strong engineering

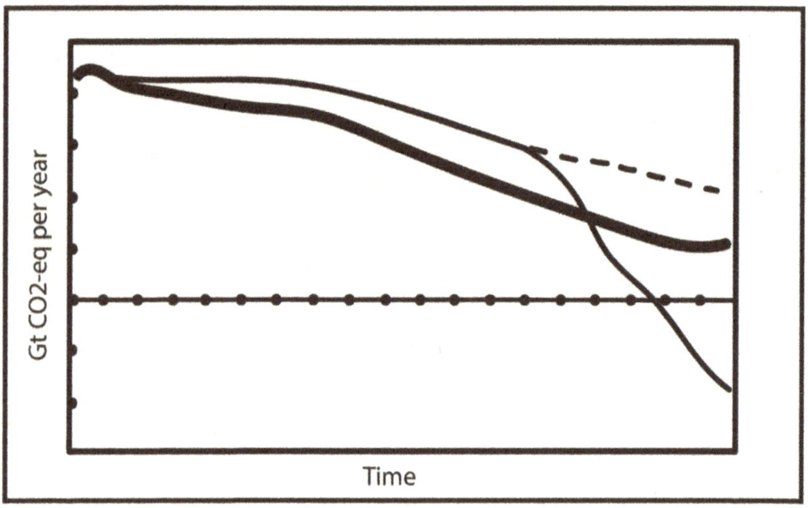

Figure 4. A simplified version of Chen and Tavoni's "Effect of DAC on GHG Abatement and Costs." The thicker line represents direct emissions reductions that would need to occur in order to reach a target of 490 ppm and keep global temperatures in check. The thin line represents what would happen to CO_2-equivalent output if current output is accompanied by increasing DAC technologies and eventually a full-scale DAC technological implementation by 2065. The dashed line represents what would occur only with some DAC implementation after 2065. The authors emphasize that while DAC could prove significant, its scale would have to be greatly increased to impact global GHGs.

resilience, showing a return to normal for global temperature within a short timeframe. Robock et al.'s modeling evidence depicts how implementing SPI could introduce a dramatic response to temperature shock. Geoengineering advocates have been keen to point out the distinctions between the horror of doing nothing, the practical, though disheartening, unlikelihood of transforming nearly all sectors of industrialized nations, and the possible results gained from aggressive forms of geoengineering like SPI. This deliberative argument is not driven by conceptions of *ecological systems theory and ecological resilience*, which would posit a distant future to which humans would need to adapt, but instead bolstered by an *engineering resilience* model, which assumes that a previous state of ecology can be achieved and that doing so quickly is the most desirable option.

The visuals in figure 4 and figure 5 portray data from empirical research and from modeling efforts, and all point toward a future where the earth will suffer from increased GHGs. While what we decide to do about these elevated levels is yet uncertain, these visuals (and others) emphasize the decreasing time in which we have to act, the improbability of existing measures to fulfill these goals, and consequently, the likelihood of geoengineering to fit into time-compressed, engineering models of resilience. Appeals to the idea of the

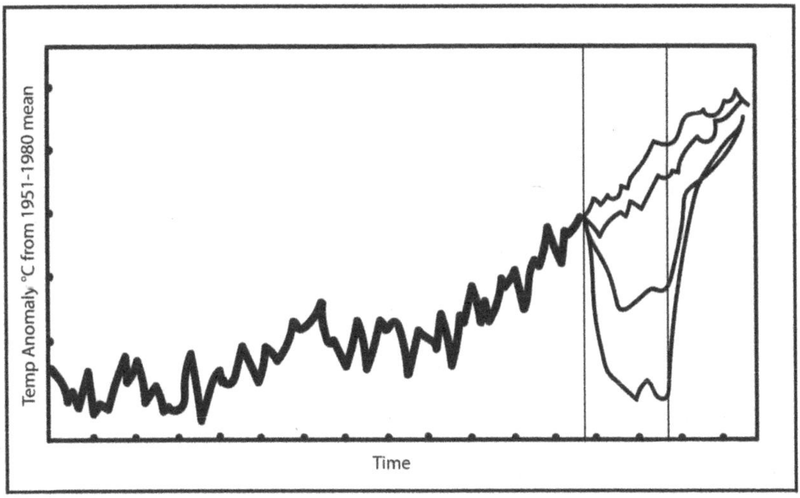

Figure 5. A simplified version of Alan Robock et al.'s representation of average temperature and geoengineering impact (from *Journal of Geophysical Research*). The thick line represents temperature up to the present day and the thinner lines represent how temperature would be affected by different intensities of SPI implementation. The vertical lines depict hypothetical starting and ending points of SPI geoengineering.

noosphere, claims about homo faber, and arguments about species-specific capacities further connect geoengineering to assumptions about humanity where humans are a hyperkeystone species able to change the conditions of its ecosystems and liable to adopt an engineering conception of resilience. This model of resiliency claims to return the earth to a previous temperature, while also emphasizing rational science as the only possible way forward—and if that response is researched thoroughly and implemented quickly, all the better. These engineering models favor SRM geoengineering over CDR, because these strategies return us from the "system shock" of global temperature change in the shortest timeframe. Frustratingly, these models also suggest that SPI geoengineering would be the most efficient, if not the safest route forward. As the following chapter shows, the assumptions about human capacity that we have seen tacitly asserted in Anthropocene and ecomodern arguments could also mask histories of Western violence and instantiate a new (Western) climactic regime.

6

GEOLOGIC VIOLENCE

As a thought experiment, imagine that a future involving some form of geoengineering is a foregone conclusion; it will be happening. What would the best version of geoengineering look like? Would it be energy-neutral, or close to it? Would it be cost-efficient, or at least show promise of future cost-reductions? Perhaps it would actively work to remove CO_2 in some fashion, thereby tackling the *actual cause* of climate change? How quickly would it take effect? Would it have widespread, if not universal, international consensus for implementation? Any proposal should probably avoid dangerous side-effects or causal ramifications that would cause further harm to humans, nonhuman animals, and ecosystems, and it should prioritize the reduction or removal of the harm of climate change from those who are most vulnerable. Planners should also work with activists, Indigenous communities, and nongovernmental organizations to ensure that all those impacted will be able to contribute to a shared governance model for the geoengineering project. Some ideas, like afforestation, regenerative agriculture, and enhanced weathering are frequently cited as fitting most of these goals. If we wanted to plant more trees, for example, Bastin et al. determined that there exists room for an additional 4.4 billion hectares of forest area on the planet. Those additional trees could store up to 205 $GtCO_2$/yr., and they found that "reaching this maximum restoration potential would reduce a considerable proportion of the global anthropogenic atmospheric carbon burden to date (~300 GtC)."[1] We could pursue tree planting regardless of its climactic benefits, but, as Holly Jean Buck has explained, when it comes to these relatively gentle CDR technologies, the problem is that they simply do not work quickly enough. "We are deluding ourselves if we think [a solution like afforestation] can be the *only* response to the disaster faced by people and species around the world," Buck argues. "But if we genuinely care about lessening climate impacts, curbing sea level rise, or saving species, other measures will be required."[2] Another much-lauded strategy with some potential for greater impact is that of DAC, a CDR geoengineering strategy which aims to pull CO_2 from the atmosphere by chemical means.

At the leading edge of DAC research has been a Canadian company, Carbon Engineering, founded by David Keith. Carbon Engineering's technology removes CO_2 from the atmosphere by putting air into contact with a sodium hydroxide-based chemical solution, resulting in small pellets of calcium carbonate that can be stored underground or, through additional chemical processes, turned into burnable fuel. Started with funds from Bill Gates and Canadian tar sands magnate Murray Edwards, in 2019 Carbon Engineering announced that it had secured an additional $68 million, mostly from fossil fuel corporations, including Chevron, BHP, and Occidental Petroleum.[3] DAC strategies are attractive to investment because they claim to provide a safe, ethical, and (eventually) profitable geoengineering method for dealing with the cause of global warming—one that sounds much less outlandish than billions of tiny space mirrors and less daunting than the wholesale and immediate reorganization of our current global energy regime. Unfortunately, for DAC to be effective at keeping temperatures down, it would need to be developed and implemented at a vast scale, something that Chen and Tavoni determined is feasible only by the latter half of this century, even if we plan for such aggressive operations now (which we are not doing). To give some sense of the scale required, we can look at Carbon Engineering's plan. Their new DAC plant should be online by 2022, at the cost of hundreds of millions of dollars. Optimistically, that plant would capture five hundred thousand metric tons of CO_2 per year, which is roughly 00.005 percent of our current, global, yearly CO_2 output. Compounding this problem of scale is the problem of finance; DAC is not yet profitable. In 2011 the American Physical Society reported that "DAC is not currently an economically viable approach to mitigating climate change," and it may take several more decades before it is cost-effective to implement.[4]

While some nongovernmental organizations like the Environmental Defense Fund have stated a preference for CDR strategies like DAC, dozens of other international organizations have suggested that no form of geoengineering is appropriate and that the "mere proposal and consideration of geoengineering as an option is already deviating attention and resources for the real alternatives to climate change."[5] Like other geoengineering strategies, DAC pulls attention and funds away from other long-term climate solutions. Greenpeace has argued that carbon capture and storage "is a costly distraction that cannot save the climate," calling it a scam intended to continue carbon industry profits.[6] It appears that some criticism of DAC is justified on technical, economic, environmental, and moral grounds; like SRM strategies, DAC fails to fully address our ingrained systems of energy production and use that have led to global warming. Instead, it currently supports existing fossil fuel energy industries and their continuing harm of global ecosystems.

The question of how geoengineering, including DAC technologies, could bring this additional harm, can be described in several ways. Certainly, the technologies themselves could cause additional environmental damage, and while I briefly comment on the dangers of DAC later in this chapter, the intricacies of these dangers are not my main concern here. Instead, I focus on two other forms of harm. First, additional efforts by science communicators to explain and describe geoengineering to a wider public could create what is known as a moral hazard, a situation where more information about a topic causes a disincentive for people to act ethically. It seems very likely that more communication about geoengineering, whatever the form described, would create a moral hazard. Second, arguments for geoengineering rely on what I call "rhetorical decoupling," when causes are argued to be decoupled from their effects. Geoengineering deliberative rhetoric uses rhetorical decoupling as means to show that economic benefits can be achieved without causing environmental destruction. Rhetorical decoupling also advances a Western paternalism and sidesteps a more inclusive deliberative process in favor of kairos-driven utilitarianism. While claiming to relieve harm to the most vulnerable populations on the planet, this rhetorical strategy could ultimately cause a different kind of harm—a violence of geologic political governance that renews long-standing power imbalances between colonial powers and those who have been colonized. Such slow violence is described by Rob Nixon in his book of the same title, as violence that is "neither spectacular nor instantaneous, but rather incremental and accretive, its calamitous repercussions playing out across a range of temporal scales."[7] Kathryn Yusoff provides the clearest articulation of this potential for ontological violence, and a lucid representation of geologic rhetoric, when she describes the attempt to impose Western, utilitarian solutions to the results of years of Western, colonial, environmental exploitation. She calls this danger the imposition of "White Geology," and I frame that possibility as the latest attempt to decouple the historical legacy of earlier geologic violence from any consequence, where decoupling rationalizes scientific policy. Accordingly, this chapter details how the problem of a moral hazard could easily add persuasive momentum to SPI geoengineering, as well as how decoupling is being turned into a rhetorical strategy that justifies a kairos-emphasized capitalism. Both situations are likely to contribute to the slow violence of geoengineering.

MORAL HAZARDS

The first form of violence this chapter considers results from a rhetorical communication problem—wherein more widely available information about geoengineering increases the likelihood that it will be pursued as a deliberative strategy. This situation is a striking representation of discursive-material persuasion, as the mere presence of information could result in material changes in the climate; it also starkly represents the presence of geosocial risk. As a moment where the geologic stratum interacts with the social stratum, the moral hazard of geoengineering clearly shows how communication could create geologic change. As the term is commonly used, a moral hazard implies that open, public discussion of a geoengineering strategy could cause a collective action problem wherein people are less motivated to take responsibility for their role in global warming and therefore care less about carbon mitigation activities, thereby increasing atmospheric CO_2 and/or advancing geoengineering.[8] After hearing of SPI, for example, people may be less dedicated to a low-carbon lifestyle or have less cause to pressure their elected officials to cut industrial carbon emissions. The concept was initially applied to situations in which obtaining insurance causes those insured to act recklessly, as an intentional consequence of having the added safety of an insurance policy. This definition shifted somewhat in the twentieth century, as analysts included unintentional actions, such as situations where installing seatbelts and airbags in cars could cause drivers to drive irresponsibly. The more generic concept of "risk compensation" is perhaps more appropriate, as it describes situations where an individual's perception of risk leads to either an increase or decrease in risk-taking behavior, regardless of the physical conditions of risk.[9] While neither term perfectly describes the condition of geoengineering (Lockley and Coffman call it a *morale* hazard, not a moral one), the terminology has stuck and has been in widespread use by geoengineering research advocates—particularly by scientists with concerns about geoengineering ethics.[10]

David Keith describes the relevance of a moral hazard for geoengineering by way of a question: "Would mere knowledge of a geoengineering method that was demonstrably low in cost and risk weaken the political will to mitigate anthropogenic climate forcing? Knowledge of geoengineering has been characterized as an insurance strategy; in analogy with the moral hazard posed by collective insurance schemes, which encourage behavior that is individually advantageous but not socially optimal, we may ascribe an analogous hazard to geoengineering if it encourages suboptimal investment in mitigation."[11] Keith's reference to "suboptimal investment" is repeated by other commentators, usually with similar definitions and concerns. The influential

2009 Royal Society report also noted these apprehensions, suggesting that widespread communication of geoengineering "could reduce the fragile political and public support for mitigation and divert resources from adaptation," although the authors also note that "there is little empirical evidence to support or refute the moral hazard argument in relation to geoengineering."[12] A more recent US Government Accountability Office report notes that "given low public awareness of geoengineering, it is difficult to determine with any confidence whether the public would reduce support for mitigation as it learned more about geoengineering."[13] The National Academy of Science echoed this concern in their geoengineering white papers, and quite frankly, the same could be said for *almost every paper in the humanities or social sciences that comments on the geoengineering ethics.*

While some critics seem uncertain as to whether there could be a moral hazard, anxiety about this appears omnipresent in the geoengineering community. A moral hazard could likely come into effect, and common sense suggests that it should at least be a worry, but published discussions tend to be split on the issue. For example, Marion Hourdequin has argued that "by diffusing attention and introducing new areas of conflict and debate, SRM geoengineering may attenuate the urgency of reducing greenhouse gases now, for the world community as a whole and for future generations."[14] Conversely, Martin Bunzl has argued that a moral hazard is unlikely, if only because he doubts that any person honestly believes that geoengineering can function as more than a temporary stopgap to real mitigation strategies. To resolve the moral hazard question, social science researchers have tested the concept through empirical study and found evidence both for it and for a "reverse moral hazard" effect, where learning about geoengineering actually increases people's active participation in carbon abatement activities—presumably because geoengineering sounds like an alarming proposition. Among those identifying a moral hazard as possible, Corner and Pidgeon found that learning about geoengineering could persuade some to take less responsibility for their carbon emissions, as "those sceptical about climate change are . . . more likely to report that they would be 'susceptible' to the trap of a personal-level moral hazard."[15] Still other studies contradict these findings, suggesting a reverse moral hazard is more likely. Wibeck et al. found that participants in their study would "consider and even advocate [for] behavioural changes, rather than being deterred by the introduction of new technologies for large-scale climate control."[16] Likewise, Merk et al. found that people were more willing to offset their own emissions when they learned about geoengineering.[17] These studies, and others, have examined whether people "believe" there is the risk for a moral hazard, and ultimately report on respondents' perceptions when asked questions about geoengineering in a controlled testing situation.

In contrast to the research discussed in the prior paragraph, studies from Kahan et al. and Fairbrother examined "whether there is a moral hazard problem," not whether people *believe* there could be a moral hazard.[18] Paradoxically, while these scholars claim to identify a reverse moral hazard, their findings also suggest additional support for geoengineering. Fairbrother asserts that his research generated "no evidence that hearing about solar radiation management affects most people's support for higher energy taxes, nor their trust in climate science"; meanwhile, for those with conservative political leanings, "hearing about SRM increased their trust in climate science."[19] His results align with Kahan et al., whose paper is perhaps the most cited study concerning the moral hazard of geoengineering. Their research study asked participants to read different news reports, then read a second report about potential environmental issues connected to rising CO_2 levels, and finally answer questions about climate change. They found that "subjects exposed to information about geoengineering *were slightly more concerned about climate change risks than those assigned to a control condition*," thereby supposedly lending evidence to the reverse moral hazard camp.[20] The problem with Kahan et al.'s work, however, is that while they confidently declare that no moral hazard would result from more public knowledge about geoengineering, they define a moral hazard as existing when that public exhibits less "concern over climate change," which is a misleading characterization.[21] A moral hazard with respect to geoengineering is not simply "less concern" about climate change but also less willingness to engage in or advocate for CO_2 mitigation. Strangely, Kahan et al. cite a range of other researchers for their definitions of a moral hazard, each of whom defines a moral hazard as less interest in mitigation and adaptation and not simply less concern about climate change.[22] This distinction is important, because additional concern about climate change—without a corresponding commitment to mitigation—could make geoengineering more desirable, not less. Despite their claimed results, Kahan et al. very likely show an increased possibility for a moral hazard in geoengineering discussions. Simply put, if more people are "concerned" about climate change but also know more about geoengineering *and* see it as a potential solution, then a moral hazard exists, regardless of their concern. Increased communication about geoengineering, I would argue, appears likely to generate additional belief in the potential for geoengineering to resolve our ongoing climate crisis, thus causing less interest in long-term, sustainable CO_2 mitigation and reduction strategies.

Ultimately, some geoengineering advocates may not be bothered by the idea of a moral hazard, in part because they feel the necessity of responding to the realities of our climate near-future and educating the public about our potential options are more important than the harm that additional public knowledge about geoengineering could cause. While Keith et al. found that a moral hazard might indeed be possible, they argued that a bigger worry would be to ignore the potential for geoengineering to counteract global warming. They claim that failing to adequately understand SRM "may well pose the greater threat to mitigation efforts, by allowing implicit reliance on SRM without scrutiny of its actual requirements, limitations and risks."[23] Even if a moral hazard is likely, some feel that such a hazard is necessary in combatting climate change. Here, the engineering definition of resilience proves to be powerful guide for this deliberative argument because it emphasizes rapidity of response above other considerations. One of the biggest challenges in geoengineering deliberation will be exactly this discursive-material communication problem—how to ensure that the very act of informing a broader public about geoengineering efforts does not also increase the likelihood that those efforts will be attempted in the future. Further, questions concerning how such information might be distributed, given the proliferation of fake news and pseudoscience, is cause for concern.[24] This potential for harm poses an ethical problem for the future of science communication—a deliberative dilemma that has no easy answer. The second form of violence that geoengineering could bring is much larger in scale than the question of a moral hazard, however. While increased information about geoengineering is liable to damage existing climate change abatement efforts, deploying geoengineering could install a new climactic regime that locks in current levels of global inequality—and this much greater potential to harm billions of people emerges from the misapplication of a rhetorical strategy to both technological and societal ends. Such a geoengineering strategy could rely on attempts to rhetorically decouple cause and effect.

RHETORICAL DECOUPLING

Decoupling is a common concept in environmental economics, one based on the idea that economic growth is causally related to (coupled with) environmental damage and degradation. When populations increase and economies expand, escalations in unsustainable agricultural practices, pollution, and GHG emissions follow that growth. These relations are thus said to be coupled. In contrast, decoupling refers to the ability of an economy or population to expand without impacting the environment at the same rates as before, and if possible, with no negative impacts at all. The United Nations Environment Programme has held a working group on large-scale decoupling prospects and in 2014 released the second of their reports on potential decoupling efforts. They see "growing evidence that decoupling will be one of the next big opportunities for economic growth, innovation and wise use of resources" and highlight different "existing technological possibilities, for developing and developed countries, and the economic advantages" of those technologies.[25] The primary avenues for this decoupling named throughout the report almost all focus on technologies that increase industrial-scale production efficiencies. The problem with decoupling, however, is not in the way it describes how energy production can be untethered from environmental destruction, but in how the concept has been used beyond this narrow definition—it has become an argument about how certain forms of cause and effect *should be* detached rather than how they *have been* detached. Here, decoupling is turned into a deliberative rhetorical strategy rather than a description of industry practice.

Over time, decoupling has been raised by geoengineering advocates and ecomodern thinkers to a rhetorical strategy par excellence. So prevalent is this argument trope throughout the geoclique that almost all geoengineering deliberations could be described as founded upon claims of decoupling. Examples of this rhetorical attempt to remove cause from effect are numerous. For some, geoengineering supposedly neutralizes, and therefore decouples, energy industries from the environmental effects of their activities by counteracting the amount of GHGs those industries release. This logic suggests that current energy regimes could be sustained well into the future, if a cost-effective, time-efficient geoengineering technology could decouple global warming from fossil fuel energy production and GHG emissions. For others, geoengineering decouples decades-long scientific research about our climate future, and the implications of that research, from the controls of established science. Subpolitical actors have based geoengineering programs upon specious science, as we have seen with ocean fertilization and SPI. Geoengineering programs might also expand upon these entrepreneurial efforts by claiming to decouple the need to respond to climactic change from state-level actors, as a new generation of subpolitical actors test their agency under ill-fitting environmental rules. This sense of decoupling further purports to decouple politics from environmental policies by providing a point of consensus that reinscribes the commitments of different groups around geoengineering concepts. Advocates for geoengineering create conditions where a strategy like SPI could be used to counteract increasing temperatures, regardless of atmospheric CO_2, and where climate change deniers can support some forms of geoengineering while disagreeing that global warming is caused by human activity. In this strategy, geoengineering may be used as a wedge to decouple politics from environmental advocacy, all while imposing a Western capitalism that represents perhaps the greatest potential source of future violence.

Ecomodern theorists also tout the advantages of decoupling economic growth from environmental damage, though this decoupling is decidedly centered around investment practices. Aggressively promoting the advantages of "accelerated, active, and conscious decoupling," the authors of "An Ecomodernist Manifesto" note that decoupling is driven by both relative decoupling (where the rate of environmental damage is slowed) and absolute decoupling (where environmental damage is halted).[26] The manifesto authors protest that such decoupling is not tied to Western capitalism, yet also posit that the drivers of decoupling are "rising demand, scarcity, inspiration, and serendipity"—essentially a romantic version of free-market capitalism.[27] While ecomodernists claim that decoupling economic growth from environmental damage does not necessarily need to take place in capitalist economies, geoengineering efforts are being most aggressively pursued in Western, neoliberal, capitalist states and have been funded by traditional extractive energy industries. The DAC technologies mentioned previously, for example, are heavily supported by the petroleum industry—a connection that exists because they represent a clear investment opportunity. These strong ties, when paired with increased public knowledge about geoengineering and an emphasis on engineering resilience, could turn decoupling into a dangerous rhetorical strategy that could be applied to areas beyond mundane industrial efficiencies. Very much like the overapplication of technological solutionism mentioned earlier, where thorny social problems are reimagined as technical problems, rhetorical decoupling could be configured as a falsely apolitical solution to resource inequality.

Ecomodernist approaches to geoengineering maintain a Western scientific worldview based upon a direct, falsely nonideological scientific pragmatism that should supposedly govern decisions about future research. This position arrives with a dangerously pragmatic ethical posture: if we value those people and cultures who will suffer the worst of climate change's effects, then we should explore geoengineering as means to avoid those impacts. Immediate, kairotic need suggests that geoengineering can be an admittedly temporary solution for addressing the climate change–caused humanitarian catastrophe that is now unfolding. However, this utilitarian solution intends to decouple historical resource inequalities and environmental violence from geologic and climactic conditions by way of a technological solutionism backed by ecomodernist entrepreneurial capitalism. The logic here is that humans could intentionally govern the climate through geoengineering and by doing so more firmly establish the geologic and atmospheric conditions for all life. At issue is how these actions could (very easily) continue a Western exploitation of disempowered peoples without the consent of those who could be most affected by geoengineering. Marion Hourdequin has maintained that geoengineering researchers like Keith have failed to demonstrate that their version of ecomodern paternalism comes with the consent of those most likely to be impacted by climate change. Keith and Horton argue that wealthy countries should conduct geoengineering research in order to assist the global poor, but Hourdequin contends that "SRM research centered in wealthy countries (as it is now) will exacerbate imbalances between powerful, wealthy countries and less powerful, less wealthy ones."[28] With this paternalistic logic, geoengineering advocates attempt to decouple the historical legacy of earlier geologic violence from any meaningful future consequences. Instead of dealing with the structural, economic inequalities caused by centuries of resource exploitation and colonial management, Western-driven geoengineering applies a technological solution that could maintain global temperatures advantageous to those same countries, while rationalizing that choice as helping to avoid a climate catastrophe *caused by those same countries*. Here, decoupling could become a rhetorical wedge used to rationalize these actions.

A WHITE GEOLOGY

Geoengineering purports to restart the world into a good Anthropocene, initiating a kind of tabula rasa for our environmental impacts by way of rhetorical decoupling. However, while the environmental impacts of new geoengineering technologies might temporarily decouple industrial-era GHG emissions from climate change, it can never decouple those emissions from the history of genocide, slavery, apartheid, colonialism, structural class and caste divisions, and environmental racism that shaped their creation. Like engineering versions of resilience, geoengineering solutions are unlikely to transform the world into an egalitarian utopia; instead, they may well restore and reinforce the geologic conditions for existing levels of inequality. In this geoengineering argument, decoupling could be applied as a deliberative strategy to a host of other social conditions, resulting in the reification of centuries of political disempowerment through intentional climactic manipulation. Kathryn Yusoff argues that any large-scale rearticulation of geologic conditions by way of geoengineering comes with the need to analyze the cultural histories that have been subsumed into *current* geologic conditions, lest we plan a future based on the racist foundations of the past. She argues that "in this moment of reinscribing geology as a property of personhood in the

Anthropocene . . . there is a need to think with its former lives of inscription, not just those currently searched for in the strata. Or, to put it another way, [we need to consider] what modes of geologic life (material and psychic) are already imbricated in geologic practices, often in violent ways."[29] Yusoff asks what geologies are being envisioned in current Anthropocene definitions and cautions that any new framing of the future must come to grips with the immoral structures upon which Western prosperity has been built. If extended to cover social conditions and cultural histories, rhetorical decoupling could conceal disempowerment and prevent it from being exposed and defined—a status quo that would be politically and economically advantageous to a neoliberal white supremacy. As Yusoff argues, "the birth of a geologic subject in the Anthropocene made without an examination of this history is a deadly erasure, rebirth without responsibility."[30] While not every advocate of geoengineering consciously promotes a rhetoric of decoupling and the erasure of history, these persuasive forces nevertheless need to be confronted in our coming deliberations. Decoupling is a dominant feature of a broader rhetoric that "creates atemporal materiality dislocated from place and time—a mythology of disassociation in the formation of matter independent of its languages of description and the historical constitution of its social relations."[31] The possibility for these rhetorics to conceal historical power imbalances is perhaps the most substantial threat they pose. Decoupling cannot be permitted to become the de facto rationale that dislocates humanity from the violences that have led to the Anthropocene, thereby concealing a history of human social relations whose effects are very much present.

We can see this Western, paternalistic, kairos-focused decoupling rhetoric at work as a core strategy of ecomodernism. In articulating their position on geoengineering, Keith and Horton have suggested that they do not believe in what has been called the "shirks responsibility" argument as it relates to geoengineering. In a "shirks responsibility" argument, the implication is that wealthy countries would be taking the easy way out by geoengineering, shirking their real responsibility by "pursuing SRM rather than a substantive programme of decarbonization."[32] Here, geoengineering circumvents the complex social and economic changes that would occur with full-scale GHG reductions, thus avoiding obligations for long-term climate justice. In their rejection of this argument, Keith and Horton appeal to kairos yet again, claiming that Western responsibilities instead lie with addressing the near-future impacts of climate change on those who will feel those effects first. They argue that research into geoengineering is imperative because the industrialized West has a responsibility to mitigate the more immediate impacts of climate change; by implication, these same countries do not have time to discuss the complex social issues connected to instituting a geoengineering plan. Because the world's most vulnerable populations are being harmed already, they argue that the real shirking of responsibility comes from how industrialized countries neglect the urgent climate situation by prioritizing overly idealistic long-term environmental goals. Those distant goals, negative carbon emissions chief among them, will do nothing in the short-term to help those without the means to adapt to rising temperatures, sea-level rise, freshwater scarcity, or widespread crop failure. While they stop short of advocating for implementation, Keith and Horton emphasize that geoengineering research is ethical because it "offers the best way to avoid these unjust [environmental and humanitarian] outcomes."[33]

The ethical stance suggested by Keith and Horton frames "traditional" (i.e., non-ecomodern) environmentalism as a Eurocentric conceit that, while understandable, ignores the immediate plight of the global poor. According to Shellenberger and Nordhaus, this environmentalism has been a failure and "must die so that something new can live."[34] They, like other ecomodern thinkers, advocate for technology and investment in order to solve environmental problems and support disengaging from attempts to save the "thing" known as the environment and instead emphasize a kairos-dominant, paternalistic postenvironmentalism. This environmentalism is echoed throughout ecomodernist thought—a hyperpractical, solution-oriented, technologically emphasized, perspective of climate change. It also goes by the name of "expertise imperialism," as Allen Buchanan has defined the term, or what Yusoff has described as an extension of "White Geology."

For Yusoff, White Geology frames the reemergence of the logics of geology in claims about the Anthropocene. It "makes legible a set of extractions, from particular subject positions, from Black and brown bodies, and from the ecologies of place" that go unstated in claims, for example, about the need to consider geoengineering as a global public good.[35] A White Geology is the continuance of Western domination and extraction practices, whereby disempowered peoples around the world are made to "benefit" from policies enacted without their consent and where their welfare is determined by those outside of their geographies and cultures. Yusoff connects these logics of domination with how the earth's resources have been extricated throughout centuries of human exploitation, and we can easily see how similar logics could be applied in geoengineering solutions. For example, in their interviews with individuals in the Solomon Islands, Alaska, and Kenya, researchers Wylie Carr and Christopher Preston found that while interview participants did not completely reject geoengineering, they voiced apprehensions about their agency in the deliberation process. People expressed fears about whether

Western states would welcome their role in a decision-making process and referenced a larger "background state of global injustice," the long history of colonialism, current global poverty and inequality, and policies of involuntary sedentarisation.[36] While only a few studies have so far assessed the anxieties of marginalized groups with respect to geoengineering, there seems to be strong concerns that those most impacted by Western-caused climate change will also be subjected to a Western-controlled geoengineering plan.[37]

As previous chapters have shown, leading geoengineering strategies have emerged from Western definitions of the Anthropocene, informed by cybernetics theories. From the very start, the idea of the Anthropocene has been infused with a supposedly rational logos, stemming from an analysis of geologic strata and increasing climactic temperature. Yet this Anthropocene also comes burdened with a deeply exploitive geologic history. Yusoff explicitly contends that the Anthropocene is the "repressed excesses" of Western geologic formations and material extractions, one that "continues to devise hegemonic aspirations that deny both those subjective repressions and the multisovereignties they represent."[38] Arguments for the potential of a "good Anthropocene," ones in which we "are as gods," are an attempt to reclaim an apolitical guiltlessness, in which a single geologic status can be applied to all of humanity.[39] The move toward species-thinking in a human-dominated era likewise silences by ignoring the long history of subjugation through a process that assumes a "colorblind" perspective on historical inequity.[40] Actions taken based upon this species-thinking are grounded on a falsely naive Western logos that emphasizes thrift, speed, and engineering resilience in the restoration of an earlier, preindustrial geologic epoch. Here, the installation of a solution supposedly free from ideology becomes another means to dispossess people from their land, but instead of a direct ejection, like an Indigenous removal policy, geoengineering could establish a new world climactic order that creates a more permanent dispossession. Yusoff argues that this attempt "to absolve the positionality of Western colonial knowledge and extraction practices, while simultaneously reinforcing and resettling them in a new territory—a Western frontier of pioneers armed with eco-optimism and geoengineering—indicates a desire to overcome coloniality without a corresponding relinquishing of the power it continues to generate in terms of who gets to formulate, implement, and speak to/of the future."[41] As the implications concerning resilience in the previous chapter suggest, geoengineering solutions are largely meant to return the global ecology to a geologic and atmospheric state that will undoubtedly reinforce the economic conditions of already industrialized, wealthy countries. Disempowered peoples and the global poor will be impacted by geoengineering proposals they may not be party to, perpetuating a logic of White Geology by way of the implementation of a "neutral" technology aligned with poorly regulated neoliberal capitalism. Such efforts will only continue the effects of slow violence that continue to echo out over centuries and cause a range of additional social and economic problems. "Slow violence," according to Nixon, "is often not just attritional but also exponential, operating as a major threat multiplier; it can fuel long-term, proliferating conflicts in situations where the conditions for sustaining life become increasingly but gradually degraded."[42] Global geoengineering is primed to become another venue for a Western imperialism, contributing to slow violence performed under the rationale of utilitarian climate change mitigation. Even in its least harmful forms, like that of DAC, current geoengineering plans carry elements of this logic.

DIRECT AIR CAPTURE

Geoengineering strategies like SPI might be the easiest to align with the nefarious logics of a White Geology, though almost all forms of climate engineering can be identified as having such potential. This includes DAC strategies—ones focused on pulling CO_2 from the air without significant risk to ecosystems—that could generate widespread public and political support, function alongside existing infrastructure, and appeal to those desiring (relatively) "environmentally friendly" solutions to climate change. Though not the geoengineering strategy capable of generating the most powerful deliberative momentum (as previous chapters have suggested), DAC represents a solution to climate change that decouples energy production from environmental side effects but also represents a slow transition to alternative energy creation.

Many DAC options involve introducing air from within the lower atmosphere to some kind of reactive sorbent, because concentrating atmospheric CO_2 from the low percentage in which it exists around us is a big challenge. This process, distinct from that which goes by the somewhat duplicitous name of "clean coal," pulls CO_2 from the air, unlike the concentrated streams of CO_2 that emerge from a powerplant. Because CO_2 occurs in ambient air at a concentration about one to three hundred times lower than that which is expelled from powerplants, carbon capture outside of those locations is much more difficult, complex, and expensive.[43] The challenge, then, is in concentrating and controlling the atmospheric CO_2 by a means that is safe, cost-efficient, and repeatable. Unfortunately, because the fledgling DAC industry needs to generate an operating budget, it is now simply functioning as another energy industry.

The biggest trial for DAC is not the chemical intricacies in capturing CO_2, however, but existing competition against other forms of fossil fuel production. Carbon Engineering, the Canadian company mentioned earlier, is trying not only to advance carbon capture technology and bring down costs but also to sustain their business in unfavorable market conditions. To become more financially viable until economics favor DAC, or until government subsidies cover the cost of extraction, DAC companies have been turning their extracted CO_2 into combustible fuel. Carbon Engineering does this by way of a process whereby a mixture of CO, CO_2, and H are chemically reacted together to produce hydrocarbons. They currently create fuel for the aerospace industry but could develop fuel for other uses at a later time. Because of this strategy, Carbon Engineering technically operates as carbon neutral. However, this use of extracted airborne CO_2 for fuel has little impact on the current amount of CO_2 in the atmosphere, as it simply replaces CO_2 that was recently in the atmosphere with CO_2 that *will soon be put back into the atmosphere* when it is consequently burned again. At the moment, Carbon Engineering's path toward becoming a sustainable energy containment company is based on the short-term plan to become another energy company, albeit a carbon-neutral one. Until there is a way for actual CO_2 sequestration to be profitable, or until the DAC industry exponentially expands, it will not start engaging in the wholesale *storage* of CO_2—something that is desperately needed if carbon capture is going to make an impact upon the current levels of CO_2 in the atmosphere. Until this happens, DAC will not substantially assist in avoiding catastrophic global temperature increases.

A second challenge for DAC companies is that they currently produce industrial waste. The most used processes in DAC are either absorption or adsorption, depending upon whether a liquid result (absorption) or a solid result (adsorption) is desired. For either adsorption or absorption, however, some kind of solvent is required, which for many companies has become sodium hydroxide (NaOH), a dangerous material also known as lye, or caustic soda. A common ingredient in commercial oven cleaners and industrial paper making, smaller amounts of NaOH can also be used to dissolve fats, produce soap, and make bagels. In a CO_2 capture plant, airborne CO_2 is introduced to NaOH, converted into sodium carbonate (Na_3CO_3), and then either precipitated out of the solution by huge quantities of water or converted into calcium carbonate ($CaCO_3$) by adding calcium hydroxide—$Ca(OH)_2$. The $CaCO_3$ is then transformed through extreme temperatures (of about 800°C) into CaO and CO_2. Companies can then use the CO_2 as they wish, often by manufacturing combustible fuel or storing it underground—and hoping that it stays there. Where this NaOH might be stored is still an open question, but it takes little imagination to suppose that it would be deposited as other hazardous chemicals or radioactive materials have been—in geographic locations that unfairly burden minorities and Indigenous groups, continuing similar practices of environmental racism that have existed for centuries.

Yet, even if these economic and technical obstacles are overcome, DAC will still have a long way to go before it can no longer be overwhelmed by the logics of White Geology. By operating within the structure of Western capitalism, and by existing as an extension of the fossil fuel industry, DAC is—at best—offering an alternative fuel source with a logic not unlike that which supports corn-based ethanol and other biofuels. Corn-based ethanol is more environmentally friendly than plain gasoline but not a solution to climate change, and its widescale implementation has destabilized the price of corn, unfairly disadvantaging those less able to cope with those changes. No geoengineering strategy will actively change global inequality unless it is accompanied by shared governance and active systems of reparation that compensate for the regimes of resource extraction that have dominated global geologic history. Geoengineering strategies that replicate existing economic inequalities are not going to do this, and neither will deliberative assertions about decoupling. Reflecting on Anthropocene and geoengineering commentaries, Yusoff explains that a future based on Western technological ingenuity will always reinforce a White Geology, because "the epochal sea change that is imagined actually reinstates the same old story of dominion, articulated through Judeo-Christian stewardship of Empire. The colonial assumption for the responsibility for and of the world is articulated anew as the white man's burden—a paternalism that is tied to a redemptive narrative of saving the world from harm on account of others while maintaining the protective thick skin of innocence."[44] Here is the rhetoric of decoupling at its most stark. The supposed goal of geoengineering is to maintain livable conditions within the critical zone of earthly habitation—the thin skin of geologic and atmospheric conditions that keep all of humanity alive. Decoupling, however, assumes that geoengineering can detach historic effects—centuries of economic inequality—from its causes: Western, capitalist resource extraction and colonial violence. Further, this deliberative decoupling assumes that the actions taken to restore and then maintain the geologic conditions of the critical zone are equivalent to an apolitical, instrumental paternalism. The "thick skin of innocence" that Yusoff names is made counterpart to the thin skin of the critical zone that geoengineering could affix in place. By naming the environmental circumstances to which we need to restore the planet and by designing geoengineering technologies to reestablish idealized critical zone conditions, geoengineering advocates attempt to decouple the recreation of geologic conditions from the political choices used to rationalize those conditions, political choices that intentionally ignore the past while proclaiming utilitarian ideals. Not only could geoengineering establish a "new normal" for the critical zone (by manufacturing "idealized" temperatures, for example)—it could instantiate those conditions as a geologic default, thereby simultaneously erasing the causes of historical inequity while reinscribing its material conditions.

The violence that geoengineering can cause is not limited to the environmental impacts of its potential implementation but also includes the damage to ongoing GHG mitigation efforts by way of a moral hazard and the dangers to climate justice efforts inherent in Western, paternalist technological solutionism. These harms are distinct from the kinds of risk described earlier, but they do have one major thing in common: both are based upon either a willful ignorance or a rejection of the geologic histories that have played a substantial role in shaping the present. Defining geoengineering's potential for violence includes evaluating how deliberative arguments for geoengineering emphasize an instrumental, overly idealistic, and falsely nonideological position. These rhetorics of decoupling are beholden to the logics of White Geology, whether in their most insidious forms, like SPI, or their most benign, like DAC. Geoengineering will likely encourage a moral hazard and thus generate its own momentum, but the greater potential for destruction comes from its capacity to mask histories of violence through actions that could cause additional violence, both slow and spectacular. In the conclusion that follows, I seek out how we might identify alternative deliberative rhetorics to arrest the spread of these discursive-material forces, though my responses are expectedly inexact and uncertain.

CONCLUSION

A Cautious Hexis

In considering how geoengineering rhetorics unfailingly emphasize technology and a false logic of rational utility, you may justifiably feel disheartened. Because so many features of geoengineering deliberations offer support for or actively encourage SPI, the implementation of this technology can seem like a foregone, albeit alarming, conclusion. The scope and intensity of these deliberative forces appear overwhelming, and like numerous climate scientists, you might believe that while geoengineering is at heart a terrible idea, and that SPI is possibly unconscionable, we might one day be forced to rely upon it. This reluctant acceptance can prove additionally exasperating because the arguments about geoengineering have been so poorly constructed. They have been duplicitous, aggressive, and worrying in equal measure and have largely taken place outside the sphere of public discussion. The climate scientists that I know are equally exasperated by how this conversation has advanced but are even more frustrated at the public's lack of sustained attention to the causes of climate change and our politicians' resistance to caring about the very clear story that climactic data is telling. "While geoengineering is awful," I heard an atmospheric scientist, and colleague, recently explain, "the effects of climate change are going to be a lot worse." Unfortunately, this is a common stance that could drown out more nuanced deliberative arguments. Many climate scientists feel as though humanity is being backed into a corner through procrastination, and they are excusably terrified of the future but aware of the potential for geoengineering to help in some rather instrumental and limited ways.

We can see the extreme of this position, and the harm it can cause, in the split in the climate activist group Extinction Rebellion (XR). Formed in the United Kingdom in 2018, XR focuses on nonviolent protests to promote awareness of climate change. When XR was taken up in the United States as XRUS, organizers created principles for effectively combatting global warming and changing our energy regime. One of their key principles demands "a just transition that prioritizes the most vulnerable people and indigenous sovereignty; establishes reparations and remediation led by and for Black people, Indigenous people, people of color and poor communities for years of environmental injustice."[1] But not all members of XR in the United States feel that we have time for such priorities. A splinter group calling itself XR America (XRA) has intentionally avoided claims about the necessity of prioritizing marginalized groups and vulnerable populations and instead emphasizes "ONE PEOPLE, ONE PLANET, ONE FUTURE."[2] As one of the founders of XRA explained, "We don't have time to argue about social justice," thereby prioritizing the immediacy of solving climate change over human rights, highlighting an instrumental kairos of engineering resilience, basing actions on Western paternalism, actively rejecting social justice and intersectional environmentalism, and unintentionally modeling a future policy on White Geology.[3] This position, and others like it, ignores how environmental damage is—and has been—disproportionately affecting marginalized people. Worst of all, it can be used by climate fascists hoping for societal and economic collapse. While most climate scientists would completely oppose climate fascism on moral and ideological grounds, their utilitarian policies can be based on similarly urgent rationales.

I am not a climate scientist, but I fully believe the scientific consensus on climate change. It is happening, we are seeing its effects now, and we are not doing enough to change the trajectory of its impact. As a rhetorician, however, I have concerns about how geoengineering deliberations have been put forth as a response to global warming. Many of the causes of climate change must be placed in context: within centuries of exploitative, Western economic policies and their intense methods of extracting value from the earth's resources. The rhetoric being used to guide our discussions about geoengineering has been falsely apolitical and often cruel. These arguments have advantageously rejected or supported the distinction between nature and culture, depending upon their objectives. The hyperpractical arguments some make for geoengineering are a utilitarian justification that could be foisted upon those with less geopolitical power and could shortcut a more open and iterative deliberative process. In part because of the lack of stasis about what climate change is and why it is happening, geoengineering feels as opportunistic as it does scientific—a resolution for a postpolitical state of exception that aligns too well with neoliberal capitalism. We have seen geoengineering ventures exploit these conditions before and are now witnessing other attempts to manipulate the coming effects of global warming. I am also deeply uncomfortable with the lineage of the rationale for geoengineering, as the major tenets of cybernetics epitomize an abstract twentieth-century view of communication, cognition, and species-thinking that should trouble modern-day rhetoricians. Cybernetics might guide utopian visions of a technological future, but like many geoengineering schemes, it falls short of being able to understand or account for the social and cultural issues that are unavoidably present. I believe that a geologic approach to rhetoric, as a broad methodology, can help us recognize and evaluate these tacit assumptions and allow rhetoricians to probe the foundations of the most prominent and powerful deliberative forces driving these environmental proposals. Yet the most significant worries I have about geoengineering come from our inability to see out of our current rhetorical, theoretical, and practical constraints. That is, we cannot at present seem to envision a convincing deliberative future from the one suggested by some geoengineering advocates, and unless we find a rhetorical approach that recognizes the long history of geologic events and their current-day impacts as rhetorical forces, *as discursive-material forces*, we may not be able to advance a more responsible geologic rhetoric.

There is no sustainable, safe, or just path forward that does not recognize the geologic, material histories that play such a large part in current forms of persuasion. Our rhetorical stances, both critical and productive, can no longer function as if the thin skin of verdant life called the critical zone is the only place where matter comes to matter. Horizontalist humanities research is an ontological cul-de-sac without geologic awareness, able only to replicate the mistakes already built into its own theoretical apparatus. A geologically attuned rhetoric tries to counter some of the conditions of horizontalist thinking and offers a place from where we can incorporate past geologic history, ontological verticality, and discursive-material rhetorics into an understanding of how persuasion occurs. I hope that this broad methodological framework will prove useful to others and shift the conversation in rhetorical studies, changing the scope and scale of the rhetoric of science and environmental rhetorics to fully account for world-altering—and potentially reality-altering—scientific endeavors.

At the same time that I am frustrated by how geoengineering has emerged and distressed by the theoretical claims of ecomodernism, I am similarly discouraged by traditional Western environmentalism, whether understood as "green" environmentalism, a romantic sense of twentieth-century environmental campaigning, or the audience of sentimental "wildlife lovers" that Mike Hulme names in *Why We Disagree about Climate Change*.[4] Though I often count myself as part of this group, what frustrates me is traditional Western environmentalism's avowed commitment to degrowth and individual performativity as the most impactful sources of change. Degrowth is a laudable goal, and despite the looming fears over postcapitalist accelerationism, should be pursued. Unfortunately, there is little actual advancement of large-scale degrowth in the United States. Thinking practically, it feels like a utopian objective, one that comes with the implausibility of all utopian concepts. Secondly, while performative environmentalist activities help, actions like sorting your garbage, carrying reusable bags, replacing your lightbulbs, buying metal straws, eating vegan, choosing not to fly, and commuting by bicycle are not going to stop the ice caps from melting, as by some accounts, we may have already reached the tipping point for their complete annihilation.[5] Performative environmentalism, including all manner of green consumerism, is not impactful at the infrastructural scale of industrial GHG emissions, and while we might be goaded into environmentally friendly behaviors through guilt, victim-blaming, pressures from the carbon industry, or commercials similar in tone to the notorious "Crying Indian" campaign, personal actions that take place within these larger industrial-level forces often do little to alter them. While painful to say, behaviors like sorting recycling matter little at the scale of climate change and geoengineering.

In order to group together the practices of environmental responsibility and performative environmentalism, environmental commentators often reference "oikonomia," or "oikos," the Greek concept for a household economy. Oikos provides the root for "ecology" and the term has been used to argue that we should think of the earth like a household and therefore practice sustainable, environmentally and economically sound principles, because we live on the earth in much the same way we live in our houses. There are good practices that can come from this (admittedly middle-class, Western) concept, yet the metaphor fails to scale. Household garbage makes up only 3 percent of the total waste in the United States, less than 10 percent of all plastic has ever been recycled, and despite how the COVID-19 pandemic impacted individual "carbon footprints," the total global output of carbon emissions is estimated to have dropped by only about 8 percent.[6] That is to say, the global output of carbon will only dip slightly during the largest societal and economic change seen in the Western world since World War II. Thinking of the planet like a house may only confuse the actions we take—and there are few actions built from this metaphor that will impact runaway climate change. Rhetorics attached to the personal and performative grounds of oikos have not "failed" any more than traditional, romantic, or green environmental rhetorics have failed, but they have been unable to operate effectively against climate change, which is why geoengineering, and proposals likely to be dominated by engineering definitions of resilience, have risen to the top of the list of potential solutions.

Along with geoengineering, how might nation-states change in response to runaway global warming? Geoff Mann and Joel Wainwright have suggested that we could be heading toward a state of exception they call a "Climate Leviathan." Named after Thomas Hobbes's description of absolute, state-level sovereign power, the Climate Leviathan "is a regulatory authority armed with democratic legitimacy, binding technical authority on scientific issues, and a panopticon-like capacity to monitor the vital granular elements of our emerging world: fresh water, carbon emissions, climate refugees, and so on."[7] In this political state, capitalism is an assumed default, climate change is a "market failure," environmental economics are implemented worldwide, and geoengineering could be an authorized solution to climate change. Under Climate Leviathan logics, the planet would be remade into a formally managed condition, where the consolidation of political and ecological control merge together, making "the state of nature and the nature of the state . . . a self-authorizing union."[8] Mann and Wainwright's envisioned outcome is perhaps preferable to that which mobilities researcher John Urry considered possible, a future "dystopian barbarism of unregulated climate change," but the neoliberal state of exception they describe would likely follow the logics of a White Geology.[9]

The most promising means of responding to runaway climate governance and aggressive geologic capitalism, Mann and Wainwright argue, is through Naomi Klein's practical suggestion to generate "Blockadia." Blockadia is Klein's term for a "roving transnational conflict zone that is cropping up with increasing frequency and intensity wherever extractive projects are attempting to dig and drill, whether for open-pit mines, or gas fracking, or tar sands oil pipelines."[10] Blockadia is not the incremental, formal policy change advocated by the lobbying groups that dominate environmentalist causes, but hyperlocal, grassroots tactics determined to disrupt fossil fuel extraction strategies as they emerge. They are often led by Indigenous groups and/or local residents, including farmers and fishers, protesting the massive extractive practices and the harm they bring to cultures and ecologies. US examples of Blockadia include shale fracking protests in Pennsylvania, mountaintop mining disputes in West Virginia, the fight over water rights in drought-stricken California, and the Dakota Access Pipeline protests at Standing Rock. These actions are those of people deeply aware of, and living with, the historic geologic effects of industry; they do not need a new geologic selfwitnessing but are actively witnessing the geologic. Blockadian actions are suffused with knowledge of the human-made geologic choices that have led to the present, and these land-oriented protests demonstrate a deep knowledge of how geosocial relations impact people. While not explicitly voicing the principles of geologic rhetoric, these confrontations are based on ungrounded, decentered rhetorical strategies and shifting tactical responses toward historical and emergent geologic injustice. Blockadia is the protest strategy of the geologic now, framed by the ethics of climate justice.

The rhetorics of Blockadia could frame a response to geoengineering, though the locations for this work are unlikely to be identifiable on a map. Geoengineering has up to this point been discussed in academic journals and taken place in remote, sometimes difficult-to-imagine sites—in the atmosphere, out at sea, or in sparsely populated wilderness. But a response to geoengineering need not be location-specific, because a reaction motivated by a geologic rhetoric is not confined to a location or even a single perspective. For Blockadia to confront geoengineering, it may need to "unground" its tactics, though such an ungrounding may conflict with existing environmental cosmologies. Startlingly, Blockadian-like responses to geoengineering may not even be situated in opposition to geoengineering. Blockadia is powered by people "driven by a desire for a deeper form of democracy, one that provides communities with real control over those resources that are most critical to collective survival—the health of the water, air, and soil," and the response to geoengineering must also adhere to these same principles.[11] That said, no geologic rhetorical strategy can prefigure a response to geoengineering. It may be easy to imagine that most groups will oppose it, but could future geoengineering plans be built from shared governance and the values of climate justice? Rather than definitively stating that such futures are impossible, a geologic rhetoric would instead support an iterative, deliberative process without foreclosing on which tactics, ideas, or commitments are acceptable. Because geoengineering does not conform to a simple yes/no ethical frame but frustratingly gives us options that were previously impossible, a response should be likewise flexible. The Harvard-based SCoPEx SPI experiment led by David Keith mentioned several times already appears to be learning these lessons the hard way.

Announced in 2015, initially planned for 2017, delayed until 2020, moved out of the United States, and then delayed again, the SCoPEx project has faced Blockadian-like opposition to its geoengineering testing ambitions. The experiments, having been moved from Arizona to New Mexico, are now planned for the small town of Kiruna, in northern Sweden, though the first tests have been canceled. Supposed to take place in June 2021, they were protested by numerous groups, including the Swedish Society for Nature Conservation, Greenpeace Sweden, Friends of the Earth Sweden, and the Indigenous Saami Council. The latter in particular noted that while they have specific environmental concerns about geoengineering, they also have deep issues with the way that the project has proceeded, noting, "We find it remarkable that the project has gone so far as to establish an agreement with SSC on test flying without, as we understand, having applied for any permits or entered into any dialogue with either the Swedish government, its authorities, the Swedish research community, Swedish civil society, or the Saami people, despite the controversial nature of SCoPEx."[12] The council also argued that they were surprised that the SCoPEx planning committee had no representation from Sweden, was composed of almost exclusively US citizens and/or residents, and appears to be "extremely homogeneous," "without any inclusion of affected groups and without directly critical and non-US voices."[13] From their perspective, no geoengineering tests should be conducted without "full, global consensus on its acceptability."[14] The tests have been rescheduled, the SCoPEx Advisory Committee is attempting to change their approach and has sent out a call for new participants, one that encourages "applications from individuals from low- and middle-income countries; women; individuals identifying as LGBTQIA+; Indigenous peoples and individuals from racialized, marginalized, colonized, and underserved groups" with backgrounds in nongovernmental research governance, equitable and inclusive societal engagement, climate change response strategies, and climate justice.[15] In this instance, then, the strategy of Blockadia has worked, at least temporarily, and seems to be nudging SCoPEx toward a more inclusive, deliberative process that includes more voices in governance.

If done well, the SCoPEx project and its permutations could serve as a foil to "experiments" like Planktos's—and serve as a template for future geoengineering projects, ones that follow a path less determined by the traditional rhetorical figure for environmental concerns, oikos, and are instead guided by a new environmental *hexis*. Hexis has been variously translated as a kind of possession, a state of having, or a disposition. Aristotle used hexis to describe the state of being in good health, and in *Metaphysics*, argued that hexis occurs between the maker and the thing made. Hexis is "a kind of activity of the haver and the had—something like an action or movement. When one thing makes and one is made, between them there is a making."[16] Between the person having a thing and the thing which is had, there is a continuously produced having. In more practical terms, this means that unlike oikos, which suggests we should be compelled to take care of the earth because it represents our home, hexis offers a different metaphor for our environmental relations, one that also recognizes the capacities that we have acquired. There can be intentional, considerate, and sustainable forms of hexis, in much the same way that there can be deliberate, thoughtful, and careful forms of oikos. Oikos gestures toward our responsibility, but many of us feel no responsibility to maintain a sustainable relationship with a world that offers us so many exploitable resources. In contrast, hexis acknowledges an active, reciprocal state of having—a situation that prompts us to ponder what we need to do about the already poor health of the world inherited by current generations. Further, it suggests that we have this responsibility, not because we are supposed to take care of our house but because we are in the act of continually producing this having. Hexis asks us to consider our disposition to what we have and what we are making of it. It asks how we want to go about continually "having" the earth. The ongoing struggle with this "having," however, is that we first need the scientific, economic, and political forces who are all too willing to consider geoengineering. We need to acknowledge that hexis means more than power over the geologic, but power with the geologic. At the risk of slipping into what Rob Nixon has called a "transcendental abstraction," a geologic hexis would support Blockadian actions by offering an ungrounded ground from which more specific local, bioregional stances could be generated.[17]

The answers to how we might continue having the earth are multiple, of course, and this book has explored and critiqued only one form—rhetorical strategies for geoengineering that often assume a "view from nowhere" or a god's eye view that believes "We are as gods and might as well get good at it." Pursuing another way of "having" is possible, though it also means accepting some of the conditions of human capacity, power, and control—something that is often difficult for those who, like myself, dislike the idea that humans should ever fundamentally change the earth. Accommodating a global hexis could, for some, take getting used to. It also may mean accepting this power at the very moment when we are realizing that there are serious, sometimes unknowable, ramifications for our actions. But this "having" is not intended to suggest an absolute jurisdiction; nor is it meant to propose that such a relationship is desirable. Instead, an active state of a continuously produced "having" means negotiating relationships between human productive capacities, material capacities, and ecosystem capacities; it means understanding the long history of causality and acting as if those causal relationships will likewise have a future; it also means bringing geology into deliberative conversations about what we could do, and should do, to sustain the relations that best serve the most vulnerable among us—with the explicit consent of those who have been and continue to be the most impacted by the geologic. The dominant guiding ethic for an alternative hexis is therefore one of climate justice, but one that does not foreclose which tactics or perspectives should be pursued.

Climate justice principles typically focus on equality, dignity, and solidarity, with an emphasis on redistributing resources, shifting away from extractive capitalism and transitioning to cooperative governance models. Many climate justice groups have argued against geoengineering. For example, the Climate Justice Alliance calls geoengineering a "false solution" technofix that avoids "the real solutions of reducing pollution at the source" and fails to "move us toward a just transition."[18] A climate justice-focused hexis could serve as a counter to the falsely apolitical, self-serving ways that geoengineering has been constructed up until now but will also need to move away from the unregulated world of subpolitical actors. It will need to recognize past geologic forces in considerations of risk, reject species-specific Anthropocene arguments, and avoid defining the conditions of climate change through economics and engineering alone. By doing so, a geologic hexis may look different for future generations—which is, after all, the ultimate goal. Kathryn Yusoff considers a similar question at the end of *A Billion Black Anthropocenes or None*: "What would a geology look like that refused its role as formulated by the adage that Charles Lyell famously coined—'the key to the past lies in the present'—and moved, instead, with the anticipatory geologic formations of the Anthropocene, a formulation that explicitly recognizes that the key to the future lies in the present and that the present is not just future oriented but that the future is already inscripted in the present, already bounding with the material recombinations of the bad material formulations of the past secreted into future possibilities?"[19] We can ask, too, about the role of rhetoric in bringing about this future geologic state of being. With a more geologically attuned approach to deliberation, we could finally believe in our unstable earth and work alongside it, within it, in all the complexities of our having, to produce a future that is livable, sustainable, and more just.

NOTES

Introduction

1. *Oxford English Dictionary Online*, s.v. "engineer," Oxford University Press, accessed October 27, 2020, www.oed.com/view/Entry/62225.

2. For more information on weather management, see James Rodger Fleming, *Fixing the Sky: The Checkered History of Weather and Climate Control* (New York: Columbia University Press, 2010) or Jacob Darwin Hamblin, *Arming Mother Nature: The Birth of Catastrophic Environmentalism* (Oxford: Oxford University Press, 2013).

3. Cesare Marchetti, "On Geoengineering and the CO_2 Problem," *Climatic Change* 1, no. 1 (1977): 59–68.

4. Royal Society, *Geoengineering the Climate: Science, Governance and Uncertainty.* (London: Royal Society, 2009), ix.

5. Clive Hamilton, *Earthmasters: The Dawn of the Age of Climate Engineering* (New Haven: Yale University Press, 2013), 51.

6. As depressing at it is, since starting this project, I have had to update this number three times, from 410 to 419. As you read, please mentally insert the new current ppm.

7. S. Self et al., "The Atmospheric Impact of the 1991 Mount Pinatubo Eruption," in *Fire and Mud: Eruptions and Lahars of Mount Pinatubo, Philippines*, ed. C. G. Newhall and R. S. Punongbayan (Seattle: University of Washington Press, 1996), 1089.

8. Steven D. Levitt and Stephen J. Dubner, *SuperFreakonomics: Global Cooling, Patriotic Prostitutes, and Why Suicide Bombers Should Buy Life Insurance* (New York: HarperCollins, 2009), 196.

9. E. Teller, L. Wood, and R. Hyde, "Global Warming and Ice Ages: I. Prospects for Physics-Based Modulation of Global Change," International Atomic Energy Agency, 1997, https://mronline.org; Stephen Salter, Graham Sortino, and John Latham, "Sea-Going Hardware for the Cloud Albedo Method for Reversing Global Warming," *Philosophical Transactions of the Royal Society A* 336, no. 1882 (November 2008): 3989–4006; Russell Seitz, "Bright Water: Hydrosols, Water Conservation and Climate Change," *Climatic Change* 105, nos. 3–4 (April 2011): 365–81; James E. Lovelock and Chris G. Rapley, "Ocean Pipes Could Help the Earth to Cure Itself," *Nature* 449, no. 403 (2007): 403.

10. Intellectual Ventures is also known as a patent troll enterprise: Morgan Baskin and Jack Denton, "'The Ultimate Patent Troll': In the Words of Technology Reporters," Pacific Standard, August 23, 2017; updated September 16, 2018, https://psmag.com; Laura Sydell and Alex Blumberg, "When Patents Attack!," This American Life, July 22, 2011, WBEZ Chicago / PRX The Public Radio Exchange, https://www.thisamericanlife.org/.

11. Erin Hale, "Geoengineering Experiment Cancelled Due to Perceived Conflict of Interest," *Guardian*, May 16, 2012, https://www.theguardian.com.

12. Andrew Yang, "It's Worse Than You Think—Lower Emissions, Higher Ground," August 18, 2020, https://2020.yang2020.com; Justine Calma, "As a Last Resort, Andrew Yang Proposes Space Mirrors to Save the Planet," *Verge*, August 26, 2019, https://www.theverge.com.

13. Aristotle, "Rhetoric," in *The Complete Works of Aristotle*, trans. and ed. Jonathan Barnes (Princeton: Princeton University Press, 1984), 1358b8–1358b20.

14. Aristotle, "Rhetoric," 1359b19–1359b23.

15. Rob Nixon, *Slow Violence and the Environmentalism of the Poor* (Cambridge, MA: Harvard University Press, 2013), 2.

16. Petra Tschakert, "Whose Hands Are Allowed at the Thermostat?" Paper presented at the conference titled "The Ethics of Geoengineering: Investigating the Moral Challenges of Solar Radiation Management," at the University of Montana, Missoula, MO, October 8–10, 2010. http://www.umt.edu.

17. Frederic Neyrat, *The Unconstructable Earth: An Ecology of Separation* (New York: Fordham University Press, 2018), 1.

18. Neyrat, *Unconstructable Earth*, 1–2.

19. Catriona McKinnon, "Sleepwalking into Lock-In? Avoiding Wrongs to Future People in the Governance of Solar Radiation Management Research," *Environmental Politics* 28, no. 3 (2019): 450.

20. McKinnon, "Sleepwalking," 448.

21. Michael Specter, "The Climate Fixers," *New Yorker*, May 7, 2012, https://www.newyorker.com.

22. James Hansen et al., "Target Atmospheric CO_2: Where Should Humanity Aim?" *Open Atmospheric Science Journal* 2, no. 1 (2008): 217–31.

23. Hansen et al., "Target Atmospheric CO_2," 11.

24. Damon H. Matthews and Ken Caldeira, "Transient Climate Carbon Simulations of Planetary Geoengineering," *PNAS* 104, no. 24 (June 2007): 1–5.

25. Jonas Anshelm and Anders Hansson, "The Last Chance to Save the Planet?: An Analysis of the Geoengineering Advocacy Discourse in the Public Debate," *Environmental Humanities* 5, no. 1 (May 2014): 107.

26. Meg Lowman, "Geo-Engineering a Rescue?" *Herald-Tribune*, August 23, 2010, https://www.heraldtribune.com.

27. Russel Berman, "The Senate Officially Believes Climate Change Is Real," *Atlantic*, January 21, 2015, https://www.theatlantic.com.

28. Suzanne Goldenberg, "US Senate Refuses to Accept Humanity's Role in Global Climate Change, Again," *Guardian*, January 22, 2015, https://www.theguardian.com.

29. Oliver Milman, "James Hansen, Father of Climate Change Awareness, Calls Paris Talks 'a Fraud,'" *Guardian*, December 12, 2015, https://www.theguardian.com.

30. Richard Heede, *Carbon Majors: Updating Activity Data, Adding Entities, and Calculating Emissions: A Training Manual* (Snowmass, CO: Climate Accountability Institute, 2019).

31. Joe Romm, "Geoengineering and the New Climate Denialism," *Think Progress*, April 29, 2009, https://thinkprogress.org.

32. William F. Lamb et al., "Discourses of Climate Delay," *Global Sustainability* 3, e17, (2020): 2.

33. Great Barrier Reef Marine Park Authority, https://www.gbrmpa.gov.au.

34. Reef Restoration and Adaptation Program, "Winning Hearts, Finding Cash, and Tough Decisions to Save the Reef," July 18, 2018, https://gbrrestoration.org.

35. Southern Cross University, "Scientists Trial World-First 'Cloud Brightening' Technique to Protect Corals," April 17, 2020, https://www.scu.edu.au.

36. Jesse Reynolds, "A Solar Geoengineering Milestone Goes Largely Unnoticed," LegalPlanet, May 6, 2020, https://legal-planet.org.

37. Tina Sikka, "A Critical Discourse Analysis of Geoengineering Advocacy," *Critical Discourse Studies* 9, no. 2 (2012): 171.

38. Bronislaw Szersynski et al. "Why Solar Radiation Management Geoengineering and Democracy Won't Mix," *Environment and Planning* 45, no. 12 (December 2013): 2814.

39. Erik Swyngedouw, "Impossible Sustainability and the Post-Political Condition," in *Making Strategies in Spatial Planning: Knowledge and Values*, ed. Maria Cerreta, Grazia Concilio, and Valeria Monno (London: Springer, 2010), 196.

40. Alain Badiou, *Metapolitics* (London: Verso, 2005).

41. Erik Swyngedouw, "The Violence of Sustainable Urbanity," *Harvard Design Magazine* 37 (Winter 2014): 25.

42. Sikka, "Critical Discourse Analysis," 167.

43. Nigel Clark, "Geoengineering and Geologic Politics," *Environment and Planning A* 45, no. 12 (December 2013): 2826.

44. Eli Kintisch, *Hack the Planet: Science's Best Hope—or Worst Nightmare—for Averting Climate Catastrophe* (Hoboken, NJ: Wiley, 2010), 195.

45. Samuel Thernstrom, "What Role for Geoengineering?" American Enterprise Institute, March 2, 2020, https://www.aei.org.

46. Sikka, 173.

47. Debra Hawhee, *Bodily Arts: Rhetoric and Athletics in Ancient Greece* (Austin, TX: University of Texas Press, 2004), 66.

48. Hawhee, *Bodily Arts*, 68.

49. Scott Porter Consigny, *Gorgias, Sophist and Artist* (Columbia: University of South Carolina Press, 2001), 87.

50. Robert P. Murphy, "The Benefits of Procrastination: The Economics of Geo-Engineering," Library of Economics and Liberty, December 7, 2009, https://www.econlib.org.

51. Robert Chris, *Systems Thinking for Geoengineering Policy: How to Reduce the Threat of Dangerous Climate Change by Embracing Uncertainty and Failure* (London: Routledge, 2015), 39.

52. For some of the numerous examples of this work, see: Thomas Rickert, "Afterword: A Crack in the Cosmic Egg, Tuning into Things," in *Rhetoric, through Everyday*

Things, ed. Scot Barnett and Casey Boyle (Tuscaloosa: University of Alabama Press, 2017), 226–32; Thomas Rickert, *Ambient Rhetoric: The Attunements of Rhetorical Being* (Pittsburgh, PA: University of Pittsburgh Press, 2013); idem., "The Whole of the Moon: Latour, Context, and the Problem of Holism," in *Thinking with Bruno Latour in Rhetoric and Composition*, ed. Paul Lynch and Nathaniel Rivers (Carbondale: Southern Illinois University Press, 2015), 135–50; Laurie E. Gries, "Iconographic Tracking: A Digital Research Method for Visual Rhetoric and Circulation Studies," *Computers and Composition* 30, no. 4 (2013): 332–48; idem., *Still Life with Rhetoric: A New Materialist Approach for Visual Rhetorics* (Logan: Utah State University Press, 2015); idem., "On Rhetorical Becoming," in Barnett and Boyle, *Rhetoric, through Everyday Things*, 155–70; Laura R. Micciche, "Writing Material," *College English* 76, no. 6 (2014): 488–505; Nathan Stormer and Bridie McGreavy, "Thinking Ecologically about Rhetoric's Ontology: Capacity, Vulnerability, and Resilience," *Philosophy and Rhetoric* 50, no. 1 (2017): 1–25; Julie Staggers and Michelle Simmons, "Secret/Agent," in *Posthuman Praxis in Technical Communication*, ed. Kristen R. Moore and Daniel P. Richards (London: Routledge, 2018), 46–66; Laurie E. Gries and Collin Gifford Brooke, *Circulation, Writing, and Rhetoric* (Logan: Utah State University Press, 2018); among many others.

53. For some examples of this work, see: Angela Haas, "Wampum as Hypertext: An American Indian Intellectual Tradition of Multimedia Theory and Practice," *Studies in American Indian Literatures* 19, no. 4 (2007): 77–100; Richard Marback, "Unclenching the Fist: Embodying Rhetoric and Giving Objects Their Due," *Rhetoric Society Quarterly* 38, no. 1 (2008): 46–65; Jeff Rice, *Networked Detroit: Rhetoric and Space in the Age of the Network* (Carbondale: Southern Illinois University Press, 2012); Jim Brown and Nathaniel Rivers, "Composing the Carpenter's Workshop," *Journal of Object Oriented Studies* 1 (2013): 27–36; Ehren Helmut Pflugfelder, "Rhetoric's New Materialism: From Micro-Rhetoric to Microbrew," *Rhetoric Society Quarterly* 45, no. 5 (2015): 441–61; Scot Barnett and Casey Boyle, eds. *Rhetoric, through Everyday Things* (Tuscaloosa: University of Alabama Press, 2016); David Grant, "Writing *Wakan*: The Lakota Pipe as Rhetorical Object," *College Composition and Communication* 69, no. 1 (2017): 61–86; among others.

54. Bridie McGreavy et al., *Tracing Rhetoric and Material Life: Ecological Approaches* (London: Palgrave Macmillan, 2018).

55. Justine Wells et al., "Introduction: Rhetoric's Ecologies," in McGreavy et al., *Tracing Rhetoric and Material Life*, 6.

56. Wells et al., "Introduction," 6–7.

57. *Cosmos: A Personal Voyage*, episode 1, "The Shores of the Cosmic Ocean," directed by Adrian Malone, Geoffrey Haines-Stiles, Rob McCain, David F. Oyster, Tom Weidlinger, and Richard J. Wells, written by Carl Sagan, Ann Druyan, and Steven Soter, hosted by Carl Sagan, aired September 28, 1980, on PBS.

58. Kathryn Yusoff, *A Billion Black Anthropocenes or None* (Minneapolis: University of Minnesota Press, 2018), 35.

59. Yusoff, *Billion Black Anthropocenes*, 35–36.

Chapter 1

1. Adam Corner and Nick Pidgeon, "Like Artificial Trees? The Effect of Framing by

Natural Analogy on Public Perceptions of Geoengineering," *Climatic Change* 130, no. 3 (June 2015): 429.

2. Matti Luokkanen, Suvi Huttunen, and Mikael Hildén, "Geoengineering, News Media and Metaphors: Framing the Controversial," *Public Understanding of Science* 23, no. 8 (November 2014): 978.

3. Luokkanen, Huttenen, and Hildén, "Geoengineering, News Media and Metaphors," 978.

4. Marilyn Strathern, "No Nature, No Culture: The Hagen Case," in *Nature, Culture and Gender*, ed. Carol MacCormack and Marilyn Strathern (Cambridge, UK: Cambridge University Press, 1980), 177. Bill McKibben later offered an alternate version of this claim, but new to this discussion are questions about how geoengineering strategies would relate to such concepts; Bill McKibben, *The End of Nature* (New York: Random House, 2006).

5. George Lakoff, "Why It Matters How We Frame the Environment," *Environmental Communication* 4, no. 1 (2010): 70–81.

6. Erle C. Ellis, "The Anthropocene: A New Epoch of Geological Time?" *Philosophical Transactions: Mathematical, Physical and Engineering Sciences* 369, no. 1938 (2011): 1027.

7. Clark, "Geoengineering and Geologic Politics," 2829.

8. C. K. Folland et al., "Observed Climate Variability and Change," in IPCC TAR 3 2001, ed. J. T. Houghton et al. (Cambridge, UK: Cambridge University Press, 2001), 140.

9. Nigel Clark, *Inhuman Nature: Sociable Life on a Dynamic Planet* (London: SAGE, 2011), 117.

10. Christopher J. Preston, *The Synthetic Age: Outdesigning Evolution, Resurrecting Species, and Reengineering Our World* (Cambridge, MA: MIT Press, 2018), 132–33.

11. Brian M. Fagan, *The Long Summer: How Climate Changed Civilization* (New York: Basic Books, 2004); Clark, *Inhuman Nature*, 117.

12. Christophe Bonneuil and Jean-Baptiste Fressoz, *The Shock of the Anthropocene: The Earth, History and Us*, trans. David Fernbach (London: Verso, 2017), 55.

13. Noel Castree and Thomas MacMillan, "Dissolving Dualisms: Actor-Networks and the Reimagination of Nature," in *Social Nature: Theory, Practice, and Politics*, ed. Noel Castree and Bruce Braun (Oxford: Blackwell Publishers, 2001), 208.

14. Carolyn Merchant, *The Death of Nature: Women, Ecology, and the Scientific Revolution* (New York: HarperCollins, 1980).

15. Merchant, *Death of Nature*, 192.

16. Merchant, 193.

17. Loraine J. Datson and Peter Galison, *Objectivity* (Cambridge, MA: MIT Press, 2010), 121.

18. Donna Haraway, "Situated Knowledges: The Science Question in Feminism and the Privilege of Partial Perspective," *Feminist Studies* 14, no. 3 (1988): 575–99.

19. Mikael Stenmark, "What Is Scientism?" *Religious Studies* 33, no. 1 (1997): 15.

20. Bonneuil and Fressoz, *Shock of the Anthropocene*, 77–78.

21. Bruno Latour, *We Have Never Been Modern*, trans. Catherine Porter (Cambridge, MA: Harvard University Press, 1993), 29.

22. Latour, *We Have Never*, 27.
23. Latour, *We Have Never*, 30.
24. Latour, *We Have Never*, 31.
25. Latour, *We Have Never*, 31.
26. Walter Mignolo, *The Darker Side of Western Modernity: Global Futures, Decolonial Options* (Durham, NC: Duke University Press, 2011), 11.
27. Mignolo, *Darker Side*, 11.
28. Mignolo, 12–13.
29. National Research Council, *Basic Research Opportunities in Earth Science* (Washington, DC: National Academies Press, 2001), 2.
30. Alexandra Arènes, Jérôme Gaillardet, and Bruno Latour, "Giving Depth to the Surface: An Exercise in the Gaia-Graphy of Critical Zones," *Anthropocene Review* 5, no. 2 (August 2018): 121.
31. Arènes, Gaillardet, and Latour, "Giving Depth to the Surface," 121.
32. Clark, *Inhuman Nature*, xvi.
33. Paul Carter, *The Lie of the Land* (London: Faber and Faber, 1996), 116; 124; Heidi Scott, "Colonialism, Landscape and the Subterranean," *Geography Compass* 2, no. 6 (November 2008).
34. Scott, "Colonialism," 1853.
35. Elizabeth Ellsworth and Jamie Kruse, "Evidence: Making a Geological Turn in Cultural Awareness," in *Making the Geologic Now: Responses to Material Conditions of Contemporary Life*, ed. Elizabeth Ellsworth and Jamie Kruse (Santa Barbara, CA: Punctum Books, 2013), 6.
36. Kathryn Yusoff, "Anthropogenesis: Origins and Endings in the Anthropocene," *Theory, Culture and Society* 33, no. 2 (March 2016): 5.
37. Yusoff, "Anthropogenesis," 4.
38. Stormer and McGreavy, "Thinking Ecologically about Rhetoric's Ontology."
39. Candice Rai and Caroline Gottschalk Druschke, eds., *Field Rhetoric: Ethnography, Ecology, and Engagement in the Places of Persuasion* (Tuscaloosa: University of Alabama Press, 2018); Amy D. Propen, *Visualizing Posthuman Conservation in the Age of the Anthropocene* (Columbus: Ohio State University Press, 2018).
40. Caroline Gottschalk Druschke, "A Tropic Future for Rhetoric Ecologies," enculturation, February 20, 2019, http://www.enculturation.net.
41. Andreas Malm and Alf Hornborg, "The Geology of Mankind? A Critique of the Anthropocene Narrative," *Anthropocene Review* 1, no. 1 (April 2014): 65.
42. Cristophe Bonneuil, "The Geological Turn: Narratives of the Anthropocene," in *The Anthropocene and the Global Environmental Crisis*, ed. Clive Hamilton et al. (London: Routledge, 2015), 20.
43. Gayatri Chakravorty Spivak, *Death of a Discipline* (New York: Columbia University Press, 2003.
44. Spivak, *Death of a Discipline*, 73.
45. Christian Moraru, *Reading for the Planet: Toward a Geomethodology* (Ann Arbor: University of Michigan Press, 2015), 6.
46. Siegfried Zielinski, *Deep Time of the Media: Toward an Archaeology of Hearing*

and Seeing by Technical Means, trans. Gloria Custance (Cambridge, MA: MIT Press, 2006); Jussi Parikka, *A Geology of Media* (Minneapolis: University of Minnesota Press, 2015).

47. Parikka, *Geology of Media*, 19.

48. Parikka, 3.

49. Parikka, 20.

50. Parikka, 4.

51. Jean-Paul Debeir, Jean-Claude Deléage, and Daniel Hémery, *In the Servitude of Power: Energy and Civilization through the Ages* (London: Zed Books, 1991).

52. Imre Szeman, *After Oil* (Morgantown: West Virginia University Press, 2016); *Energy Culture: Art and Theory on Oil and Beyond* (Morgantown: West Virginia University Press, 2019); *On Petrocultures: Globalization, Culture, Energy* (Morgantown: West Virginia University Press, 2019).

53. Sheena Wilson, Adam Carlson, and Imre Szeman, "On Petrocultures: Or, Why We Need to Understand Oil to Understand Everything Else," in *Petrocultures: Oil, Politics, Culture*, ed. Sheena Wilson, Adam Carlson, and Imre Szeman (Montreal: McGill-Queen's University Press, 2017).

54. Wilson, Carlson, and Szeman, "On Petrocultures," 3.

55. Imre Szeman and Dominic Boyer, "Introduction: On the Energy Humanities," in *Energy Humanities: An Anthology*, ed. Imre Szeman and Dominic Boyer (Baltimore: Johns Hopkins University Press, 2017), 36.

56. Wilson, Carlson, and Szeman, 4.

57. William E. Connolly, *Climate Machines, Fascist Drives, and Truth* (Durham, NC: Duke University Press, 2019), 42.

58. Connolly, *Climate Machines*, 42.

59. Connolly, 40.

60. Clark, *Inhuman Nature*, xvii–xviii.

61. Kathryn Yusoff, "The Geoengine: Geoengineering and the Geopolitics of Planetary Modification," *Environment and Planning A: Economy and Space* 45, no. 12 (December 2013): 2806. The Edwards work that Yusoff cites here is Paul N. Edwards, *A Vast Machine: Computer Models, Climate Data, and the Politics of Global Warming* (Cambridge, MA: MIT Press, 2010), 145.

62. Yusoff, "Geoengine," 2800.

63. Jeanne Fahnestock, "Rhetoric of Science: Enriching the Discipline," *Technical Communication Quarterly* 14, no. 3 (2005): 278.

64. Fahnestock, "Rhetoric of Science," 278.

65. David J. Depew and John Lyne, "The Productivity of Scientific Rhetoric," *Poroi* 9, no. 1 (2013): 2.

66. Kenny Walker, Jennifer Malkowsky, and Damien Smith Pfister, "A Choreography of Living Texts: Selections from the ARST Oral History Project," *Rhetoric Review* 33, no. 3 (2014): 27.

67. Alan G. Gross, "Rhetoric of Science," in *Encyclopedia of Rhetoric and Composition*, ed. Theresa Enos (London: Routledge, 1996), 622.

68. Chaïm Perelman and L. Olbrechts-Tyteca, *The New Rhetoric: A Treatise on*

Argumentation, trans. John Wilkinson and Purcell Weaver (South Bend, IN: University of Notre Dame Press, 1973).

69. Some readers will notice that much of the theoretical grounding of this book is based on the work of sociological theorists. In some ways, the approach I take here is simply an updating of the sociological ground for a rhetoric of science and the environment.

70. Steve Fuller, "The Strong Program in the Rhetoric of Science," in *Science, Reason and Rhetoric*, ed. Henry Krips, J. E. McGuire, and Trevor Melia (Pittsburgh, PA: University of Pittsburgh Press, 1995), 110.

71. Ken Baake, *Metaphor and Knowledge: The Challenges of Writing Science* (Albany: SUNY Press, 2003), 45.

72. Randy Allen Harris, "Knowing, Rhetoric, Science," in *Visions and Revisions: Continuity and Change in Rhetoric and Composition*, ed. James D. Williams (Carbondale: Southern Illinois University Press, 2002), 170–71.

73. Barry Barnes, *Scientific Knowledge and Sociological Theory* (London: Routledge and K. Paul, 1974); David Bloor, *Knowledge and Social Imagery* (London: Routledge, 1976).

74. Alan G. Gross, "Rhetoric of Science 'Is' Epistemic Rhetoric," *Quarterly Journal of Speech* 76, no. 3 (1990): 304–6.

75. Jeanne Fahnestock, "The Rhetorical Arts of Cooperation," *JGE: The Journal of General Education* 62, no. 1 (2013): 14.

76. Carl G. Herndl and Lauren Leigh Cutlip, "'How Can We Act?' A Praxiographical Program for the Rhetoric of Technology, Science, and Medicine," *Poroi* 9, no. 1 (2013): 2.

77. David R. Gruber and Randy Allen Harris, "Scientific Futures for a Rhetoric of Science: 'We Do This and They Do That?' A Junior-Senior Scholar Session, RSA 2018, Minneapolis, Minnesota, USA; 1 June 2018," *Poroi* 14, no. 2 (2019): 4–5.

78. Rickert, *Ambient Rhetoric*, 194.

79. Byron Hawk, *A Counter-History of Composition: Toward Methodologies of Complexity* (Pittsburgh, PA: University of Pittsburgh Press, 2007).

80. Parmenides of Elea, *Fragments*, trans. David Gallop (Toronto: University of Toronto Press, 2013).

81. Wilhelm Joseph von Schelling, *Ideas for a Philosophy of Nature*, trans. Errol E. Harris and Peter Heath (Cambridge, UK: Cambridge University Press, 2001), 187.

82. Heraclitus is perhaps best known for stating that it is "impossible to ever step into the same river twice," though that precise phrase never appears in his works. Jonathan Barnes, in *The Presocratic Philosophers*, notes that Heraclitus's meaning is occluded by time but suggests that the obvious reading is that "stability may cover constant change." Barnes, *The Presocratic Philosophers* (London: Routledge, 1982), 50.

83. Wilhelm Joseph von Schelling, *First Outline of a System of the Philosophy of Nature*, trans. Keith R. Peterson (Albany: SUNY Press, 2004), 289.

84. Hawk, *Counter-History*, 142.

85. Hawk, 142.

86. Diana Coole, "The Inertia of Matter and the Generativity of Flesh," in *New*

Materialisms: Ontology, Agency, and Politics, ed. Diana Coole and Samantha Frost (Durham, NC: Duke University Press, 2010), 99.

87. Hawk, 140.

88. Henri Bergson, *Creative Evolution*, trans. Arthur Mitchell (New York: Henry Holt, 1911), 204.

89. Bergson, *Creative Evolution*, 14. Emphasis in original.

90. Iris Van der Tuin, "A Different Starting Point, a Different Metaphysics: Reading Bergson and Barad Diffractively," *Hypatia* 26, no. 1 (2011): 22–42.

91. Karen Michelle Barad, *Meeting the Universe Halfway: Quantum Physics and the Entanglement of Matter and Meaning* (Durham, NC: Duke University Press, 2007).

92. Barad, *Meeting the Universe*, 151.

93. Barad, 151.

94. Barad, 140.

95. Barad, 91.

96. Barad, 138.

97. Barad, 153.

98. Jerry Lee Rosiek, Jimmy Snyder, and Scott L. Pratt, "The New Materialisms and Indigenous Theories of Non-Human Agency: Making the Case for Respectful Anti-Colonial Engagement," *Qualitative Inquiry* 26, nos. 3–4 (March–April 2020): 331.

99. Rosiek, Snyder, and Pratt, "New Materialisms," 337.

100. Rosiek, Snyder, and Pratt, 337.

101. Gabriela Raquel Rìos, "Cultivating Land-Based Literacies and Rhetorics," *Literacy in Composition Studies* 3, no. 1 (2015): 64.

102. Kim TallBear, "Beyond the Life/Not Life Binary: A Feminist-Indigenous Reading of Cryopreservation, Interspecies Thinking and the New Materialisms" in *Cryopolitics: Frozen Life in a Melting World*, ed. Joanna Radin and Emma Kowal (Cambridge, MA: MIT Press, 2017), 198.

103. Vanessa Watts, "Indigenous Place-Thought and Agency amongst Humans and Non-Humans (First Woman and Sky Woman Go on a European World Tour!)," *Decolonization: Indigeneity, Education & Society* 2, no. 1 (2013): 26.

104. Watts, "Indigenous Place-Thought," 25.

105. TallBear, "Beyond the Life/Not Life Binary." 191.

106. Mel Y. Chen, *Animacies: Biopolitics, Racial Mattering, and Queer Affect* (Durham, NC: Duke University Press, 2012), 237.

107. Robin Wall Kimmerer, *Braiding Sweetgrass: Indigenous Wisdom, Scientific Knowledge and the Teachings of Plants* (Minneapolis: Milkweed Editions, 2013), 54–55.

108. Scott, "Colonialism," 1854.

109. Timothy Morton, *The Ecological Thought* (Cambridge, MA: Harvard University Press, 2010), 31.

110. Yusoff, "Anthropogenesis," 7. Emphasis mine.

111. Gilles Deleuze and Félix Guattari, *A Thousand Plateaus: Capitalism and Schizophrenia*, trans. Brian Massumi (Minneapolis: University of Minnesota Press, 1987), 40.

112. Mark Bonta and John Protevi, *Deleuze and Geophilosophy: A Guide and Glossary* (Edinburgh: Edinburgh University Press, 2004), 4, 14.

113. Deleuze and Guattari define "strata" and "assemblage" differently, seeing a state-level apparatus or tightly knit social group as "strata" and more loosely held "sympathies" as "assemblages." I avoid to parsing out these differences in detail here but direct readers to Manuel DeLanda's work, especially his chapter "Deleuzian Social Ontology and Assemblage Theory," in *Deleuze and the Social*, ed. Martin Fuglsang and Bent Meier Sorensen (Edinburgh: Edinburgh University Press, 2006).

114. Kathryn Yusoff, "Geosocial Strata," *Theory, Culture & Society* 34, nos. 2–3 (March–May 2017): 105.

115. Yusoff, "Geosocial Strata," 108.

116. Deleuze and Guattari, *Thousand Plateaus*, 335.

117. Yusoff, "Geosocial Strata," 110–11.

118. Deleuze and Guattari, 335.

119. Clark, *Inhuman Nature*, 46–47.

120. Oxford English Dictionary Online, s.v. "primordial," accessed June 12, 2020, www.oed.com.

121. I would be remiss if I failed to note that many Indigenous land-based ontologies seem to take this claim as a given.

122. Clark, *Inhuman Nature*, 19–20.

123. Clark, *Inhuman Nature*, 20.

124. Yusoff, "Geosocial Strata," 117.

125. Eugene W. Holland, *Deleuze and Guattari's* A Thousand Plateaus (London: Bloomsbury, 2013), 65.

126. Nigel Clark, "Politics of Strata," *Theory, Culture and Society* 34, nos. 2–3 (March–May 2017): 213.

127. Deleuze and Guattari, 63.

128. Nathaniel A. Rivers, "Deep Ambivalence and Wild Objects: Toward a Strange Environmental Rhetoric," *Rhetoric Society Quarterly* 45, no. 5 (2015): 422.

129. Propen, *Visualizing Posthuman Conservation*, 4.

130. Jennifer Clary-Lemon, *Planting the Anthropocene: Rhetorics of Natureculture* (Logan: Utah State University Press, 2019), 59.

131. Clary-Lemon, *Planting the Anthropocene*, 59

132. Clark, "Politics of Strata," 228.

Chapter 2

1. House of Commons Science and Technology Committee, "The Regulation of Geoengineering," The Stationery Office Limited House of 23, Fifth Report of Session, 2009–2010, March 10, 2010, 23, https://publications.parliament.uk; Yuri Izrael et al., "Field Experiment on Studying Solar Radiation Passing through Aerosol Layers," *Russian Meteorology and Hydrology* 34, no. 5 (2009): 272.

2. Izrael et al., "Field Experiment," 272.

3. Yuri Izrael, Alexey Ryaboshapko, and N. N. Petrov, "Comparative Analysis of Geo-Engineering Approaches to Climate Stabilization," *Russian Meteorology and Hydrology* 34, no. 6 (2009): 345.

4. Leah Ceccarelli, "Manufactured Scientific Controversy: Science, Rhetoric, and Public Debate," *Rhetoric and Public Affairs* 14, no. 2 (Summer 2011): 197.

5. Anthony Leiserowitz et al., "Climate Change in the American Mind: December 2018," Yale Program on Climate Change Communication and George Mason University Center for Climate Change Communication, January 22, 2019, https://climatecommunication.yale.edu.

6. George A. Kennedy, *A New History of Classical Rhetoric* (Princeton, NJ: Princeton University Press, 1994), 98.

7. Cicero, *Topica*, trans. Tobias Reinhardt (Oxford: Oxford University Press, 2003); Lawrence J. Prelli, "Stasis and the Problem of Incommensurate Communication: The Case of Spousal Violence Research," in *Rhetoric and Incommensurability*, ed. Randy Allen Harris (West Lafayette, IN: Parlor Press, 2005), 300.

8. Michael E. Mann, *The Hockey Stick and the Climate Wars: Dispatches from the Front Lines* (New York: Columbia University Press, 2012), 23.

9. John Law, *After Method: Mess in Social Science Research* (London: Routledge, 2004), 159.

10. John Law, "Enacting Naturecultures: A View from STS," *Heterogeneities*, February 24, 2004, www.heterogeneities.net, 5.

11. Law, "Enacting Naturecultures," 5.

12. Annemarie Mol, "Ontological Politics: A Word and Some Questions," Supplement, *Sociological Review* 47, no. 1 (1999): 77.

13. Annemarie Mol, *The Body Multiple: Ontology in Medical Practice* (Durham, NC: Duke University Press, 2003), vii–viii.

14. Mol, *The Body Multiple*, 5.

15. Luigi Pellizzoni, *Ontological Politics in a Disposable World: The New Mastery of Nature* (London: Routledge, 2015), 86.

16. Bruno Latour, *Pandora's Hope: Essays on the Reality of Science Studies* (Cambridge, MA: Harvard University Press, 1999), 155–56.

17. Latour, *Pandora's Hope*, 148.

18. Latour, *Pandora's Hope*, 173.

19. Gilles Deleuze and Claire Parnet, *Dialogues*, trans. Hugh Tomlinson and Barbara Habberjam (Paris: Flammarion, 1977), 69.

20. Law, *After Method*, 42.

21. S. Scott Graham and Carl Herndl, "Multiple Ontologies in Pain Management: Toward a Postplural Rhetoric of Science," *Technical Communication Quarterly* 22, no. 2 (2013): 113.

22. Sean Esbjörn-Hargens, "An Ontology of Climate Change: Integral Pluralism and the Enactment of Multiple Objects," *Journal of Integral Theory and Practice* 5, no. 1 (2010): 143–74.

23. Svante Arrhenius, "On the Influence of Carbonic Acid in the Air upon the Temperature of the Ground," *London, Edinburgh, and Dublin Philosophical Magazine and Journal of Science* 41, no. 251 (1896): 266.

24. G. S. Callendar, "The Artificial Production of Carbon Dioxide and Its Influence

on Temperature," *Quarterly Journal of the Royal Meteorological Society* 64, no. 275 (1938): 223–40.

25. Spencer Weart, "The Carbon Dioxide Greenhouse Effect," Discovery of Global Warming, www.aip.org.

26. Norris W. Rakestraw and Charles D. Keeling, "Proposal to the National Science Foundation for A Study of the Abundance of CO_2," Scripps Institution of Oceanography, Folder "Government Correspondence—NSF - 1958–1969," Keeling office files, SIO, June 1958, 38.

27. Pieter Tans and Kirk Thoning, "How We Measure Background CO_2 Levels on Mauna Loa," NOAA Earth System Research Laboratory, https://gml.noaa.gov.

28. Edwards, *A Vast Machine*, 145.

29. Edwards, 342–46.

30. Edwards, 346.

31. Edwards, 342.

32. Rickert, *Ambient Rhetoric*, 8–9.

33. Aimee Kendall Roundtree, *Computer Simulation, Rhetoric, and the Scientific Imagination: How Virtual Evidence Shapes Science in the Making and in the News* (Lanham, MD: Lexington Books, 2017), 15, 12.

34. Naomi Oreskes, Kristin Shrader-Frenchette, and Kenneth Belitz, "Verification, Validation, and Confirmation of Numerical Models in the Earth Sciences," *Science* 263, no. 5147 (1994): 644.

35. For much more detail on the history and development of environmental economics, see Agnar Sandmo, "The Early History of Environmental Economics," *Review of Environmental Economics and Policy* 9, no. 1 (Winter 2015): 43–63.

36. Kenneth E. Boulding, "The Economics of the Coming Spaceship Earth," in *Environmental Quality in a Growing Economy: Essays from the Sixth RFF Forum*, ed. H. Jarrett, (Baltimore: Johns Hopkins University Press, 1966), 3–14.

37. Boulding, "The Economics," 9–10.

38. A. Denny Ellerman et al., *Markets for Clean Air: The Acid Rain Program* (Cambridge, UK: Cambridge University Press, 2000).

39. Jun Li, "Governing Urban Infrastructure in Developing Cities: The Role of Carbon Finance," in *Global Change, Energy Issues and Regulation Policies*, ed. Jean Bernard Saulnier and Marcelo D. Varella (Heidelberg: Springer, 2013), 262.

40. Pellizzoni, *Ontological Politics*, 17.

41. Pellizzoni, 81.

42. Pellizzoni, 81.

43. Kelly Levin, "New Global CO_2 Emissions Numbers Are In. They're Not Good," *World Resources Institute*, December 5, 2018, https://www.wri.org.

44. Intergovernmental Panel on Climate Change, *Special Report: Global Warming of 1.5°C*, https://www.ipcc.ch, 6; 14.

45. Interagency Working Group on Social Cost of Carbon, "Technical Support Document: Social Cost of Carbon for Regulatory Impact Analysis under Executive Order 12866," United States Government, 2010, www.epa.gov.

46. Peter J. Jacques, Riley E. Dunlap, and Mark Freeman, "The Organisation of

Denial: Conservative Think Tanks and Environmental Scepticism," *Environmental Politics* 17, no. 3 (2008): 354.

47. Intergovernmental Panel on Climate Change, *Climate Change 2014 Mitigation of Climate Change, The Working Group III Contribution to the Fifth Assessment Report (AR5) of the Intergovernmental Panel on Climate Change* (Cambridge, UK: Cambridge University Press, 2014), 41.

48. Daniel Cressey, "Latest IPCC Climate Report Puts Geoengineering in the Spotlight," *Scientific American*, October 2, 2013, https://www.scientificamerican.com.

49. Miranda Boettcher and Stefan Schäfer, "Reflecting Upon 10 Years of Geoengineering Research: Introduction to the Crutzen + 10 Special Issue," *Earth's Future* 5, no. 3 (March 2017): 268.

50. Alan Robock et al., "Benefits, Risks, and Costs of Stratospheric Geoengineering," *Geophysical Research Letters* 36, no. 19 (October 2009).

51. Graham and Herndl, "Multiple Ontologies," 123.

52. Dozens of other researchers have explored this ground before. For some excellent examples, see: Mark Hoofnagel, "Climate Change Deniers: Failsafe Tips on How to Spot Them," *Guardian*, March 11, 2019, https://www.theguardian.com; Naomi Oreskes and Erik M. Conway, *Merchants of Doubt: How a Handful of Scientists Obscured the Truth on Issues from Tobacco Smoke to Global Warming* (New York: Bloomsbury Press, 2010); Robert J. Brulle, "Institutionalizing Delay: Foundation Funding and the Creation of Climate Change Counter-Movement Organizations," *Climatic Change* 122, no. 4 (February 2014): 681–94.

53. R. E. Dunlap and A. M. McCright, "Organized Climate Change Denial," in *The Oxford Handbook of Climate Change and Society* (Oxford: Oxford University Press, 2011), 144–60.

54. Levitt and Dubner, *SuperFreakonomics*, 169.

55. Thernstrom, "What Role for Geoengineering?"

56. J. Eric Bickel and Lee Lane, "An Analysis of Climate Engineering as a Response to Climate Change," Copenhagen Consensus Center, 2009, copenhagenconsensus.com, 5.

57. Richard A. Epstein, "A Climate Change Emergency?" Hoover Institution, October 7, 2019, https://www.hoover.org; Fred Iklé and Lowell Wood, "Sunscreen for Planet Earth," *Hoover Digest* 1, 2008, https://www.hoover.org.

58. Iklé and Wood, "Sunscreen for Planet Earth."

59. Brian Kahn, "Republicans Decided to Talk about a Dramatic Climate Solution without Addressing the Actual Problem," *Gizmodo*, August 11, 2017, https://gizmodo.com.

60. Andy Biggs, "Congressman Andy Biggs' Opening Statement on Joint Subcommittee Hearing on Geoengineering," November 8, 2017, https://biggs.house.gov.

61. Bickel and Lane, "Analysis," 46.

62. Philip Rasch, Simone Tilmes, Richard P. Turco, et al., "An Overview of Geoengineering of Climate using Stratospheric Sulfate Aerosols," *Philosophical Transactions of the Royal Society A* 366, no. 1882 (2008): 4007–37.

63. Naomi E. Vaughan and Timothy M. Lenton, "A Review of Climate Geoengineering Proposals," *Climatic Change* 109, nos. 3–4 (December 2011): 765.

64. Vaughan and Lenton, "Review," 782.

65. Marlos Goes, Nancy Tuana, and Klaus Keller, "The Economics (or Lack Thereof) of Aerosol Geoengineering," *Climatic Change* 109, nos. 3–4 (December 2011): 720.

66. House Select Committee on the Climate Crisis, *Solving the Climate Crisis: The Congressional Action Plan for Clean Energy and a Healthy, Resilient, and Just America*, June 2020, https://climatecrisis.house.gov, 526.

Chapter 3

1. Dave Levitan, "For Geoengineers, a Scientific Existential Crisis," UNDARK, January 1, 2019, https://undark.org.

2. Alan Robock, "20 Reasons Why Geoengineering May Be a Bad Idea," *Bulletin of the Atomic Scientists* 64, no. 2 (2008): 16.

3. These conclusions have been reached by dozens of independent researchers, though for an overview of these dangers, see Vaughan and Lenton, "Review"; Alan Robock et al., "Studying Geoengineering with Natural and Anthropogenic Analogs," in "Geoengineering Research and Its Limitations," Special issue, *Climatic Change* 121, no. 3 (December 2013): 445–58; and Rob Bellamy et al., "A Review of Climate Geoengineering Appraisals," *Wiley Interdisciplinary Reviews: Climate Change* 3, no. 6 (November–December 2012): 597–615.

4. C. S. Zerefos et al., "Atmospheric Effects of Volcanic Eruptions as Seen by Famous Artists and Depicted in Their Paintings," *Atmospheric Chemistry and Physics* 7, no. 15 (2007).

5. Paul Slovic and Elke U. Weber, "Perception of Risk Posed by Extreme Events," paper presented at the conference "Risk Management Strategies in an Uncertain World," Palisades, New York, April 12–13, 2002, 4.

6. Sandrine Tranchard, "The New ISO 31000 Keeps Risk Management Simple," International Organization for Standardization, February 15, 2018, https://www.iso.org.

7. Sue McKell and Paul De Barro, "Between Two Absolutes Lies Risk: Risk Communication in Biosecurity Discourse," in *Communicating Risk*, ed. Jonathan Crichton, Christopher N. Candlin, and Arthur S. Firkins (London: Palgrave Macmillan, 2016), 232.

8. William Leiss and Douglas Powell, *Mad Cows and Mother's Milk: The Perils of Poor Risk Communication* (Montreal: McGill-Queen's University Press, 1997), 26.

9. Izrael, Ryaboshapko, and Petrov, "Comparative Analysis," 344.

10. Izrael, Ryaboshapko, and Petrov, 344.

11. Angus J. Ferraro, Andrew J. Charlton-Perez, and Eleanor J. Highwood, "A Risk-Based Framework for Assessing the Effectiveness of Stratospheric Aerosol Geoengineering," *PLoS ONE* 9, no. 2 (2014): 6.

12. Suvi Huttunen and Mikael Hildén, "Framing the Controversial: Geoengineering in Academic Literature," *Science Communication* 36, no. 1 (February 2014): 11.

13. Philip W. Boyd, "Ranking Geo-Engineering Schemes," *Nature Geoscience* 1, no. 11 (2008): 723–24.

14. Boyd, "Ranking Geo-Engineering," 723–24.

15. Bellamy et al., "Review of Climate Geoengineering Appraisals," 606.

16. Sikka, "Critical Discourse Analysis."

17. Steven. B. Katz and Carolyn R. Miller, "The Low-Level Radioactive Waste Siting Controversy in North Carolina: Toward a Rhetorical Model of Risk Communication," in *Green Culture: Environmental Rhetoric in Contemporary America*, ed. C. G. Herndl and S. C. Brown (Madison: University of Wisconsin Press, 1996), 116.

18. Jeffrey T. Grabill and W. Michele Simmons, "Toward a Critical Rhetoric of Risk Communication: Producing Citizens and the Role of Technical Communicators," *Technical Communication Quarterly* 7, no. 4 (1998): 419.

19. Grabill and Simmons, "Toward a Critical Rhetoric," 429.

20. Mary Douglas and Aaron B. Wildavsky, *Risk and Culture: An Essay on the Selection of Technical and Environmental Dangers* (Berkeley: University of California Press, 1982).

21. Philippe Hanna, Frank Vanclay, and Jos Arts, "The Communication and Management of Social Risks and Their Relevance to Corporate-Community Relationships," in Crichton, Candlin, and Firkins, *Communicating Risk*, 172.

22. Karen L. Henwood and Nick Pidgeon, "Interpretive Environmental Risk Research: Affect, Discourses and Change," in Crichton, Candlin, and Firkins, 157.

23. Beverly A. Sauer, *The Rhetoric of Risk: Technical Documentation in Hazardous Environments* (London: Routledge, 2002); W. Michele Simmons, *Participation and Power: Civic Discourse in Environmental Policy Decisions* (Albany: SUNY Press, 2007); Erin A. Frost, "Transcultural Risk Communication on Dauphin Island: An Analysis of Ironically Located Responses to the Deepwater Horizon Disaster," *Technical Communication Quarterly* 22, no. 1 (2013): 50–66; Liza Potts, *Social Media in Disaster Response: How Experience Architects Can Build for Participation* (London: Routledge, 2014).

24. Adam Corner and Nick Pidgeon, "Geoengineering, Climate Change Scepticism and the 'Moral Hazard' Argument: An Experimental Study of UK Public Perceptions," *Philosophical Transactions. Series A* 372, no. 2031 (2014): 428.

25. Dirk Scheer and Ortwin Renn, "Public Perception of Geoengineering and Its Consequences for Public Debate," *Climatic Change* 125, nos. 3–4 (August 2014): 308.

26. Sheila Jasanoff and Sang-Hyun Kim, "Containing the Atom: Sociotechnical Imaginaries and Nuclear Power in the United States and South Korea," *Minerva* 47, no. 2 (June 2009): 120.

27. B. Sheppard, M. Janoske, and B. Liu, *Understanding Risk Communication Theory: A Guide for Emergency Managers and Communicators*, Report to Human Factors/Behavioral Sciences Division, Science and Technology Directorate, US Department of Homeland Security. College Park, MD, 2012, 6.

28. Sheppard, Janoske, and Liu, *Understanding Risk*, 6.

29. Ulrich Beck, *Risk Society: Towards a New Modernity*, trans. Mark Ritter (London: Sage, 1992).

30. Ulrich Beck, "The Reinvention of Politics: Towards a Theory of Reflexive Modernization," in *Reflexive Modernization: Politics, Tradition and Aesthetics in the Modern Social Order*, ed. Ulrich Beck, Anthony Giddens, and Scott Lash (Stanford, CA: Stanford University Press, 1994), 6.

31. Beck, *Risk Society*, 14.

32. Beck, "The Reinvention of Politics," 6.

33. Beck, *Risk Society*, 19.

34. Beck, "The Reinvention of Politics," 45.

35. J. Blake Scott, "Kairos as Indeterminate Risk Management: The Pharmaceutical Industry's Response to Bioterrorism," *Quarterly Journal of Speech* 92, no. 2 (2006): 119.

36. Robert Danisch, "Political Rhetoric in a World Risk Society," *Rhetoric Society Quarterly* 40, no. 2 (2010): 173–74.

37. Beck, *Risk Society*, 13.

38. Danisch, "Political Rhetoric," 178.

39. Deborah Lupton, *Risk* (London: Routledge, 1999), 64.

40. Paul Crutzen, "Geology of Mankind," *Nature* 415, no. 6867 (2002).

41. Dipesh Chakrabarty, "The Climate of History: Four Theses," *Critical Inquiry* 35, no. 2 (Winter 2009): 208.

42. Stephen Gardiner, *A Perfect Moral Storm: The Ethical Tragedy of Climate Change* (Oxford: Oxford University Press, 2013), 33.

43. Beck, *Risk Society*, 21.

44. Andreas Malm, *Fossil Capital: The Rise of Steam Power and the Roots of Global Warming* (London: Verso, 2016), 8.

45. Ulrich Beck, "Incalculable Futures: World Risk Society and Its Social and Political Implications," in *Ulrich Beck: Pioneer in Cosmopolitan Sociology and Risk Society*, ed. Ulrich Beck (Heidelberg: Springer, 2014), 82.

46. Nigel Clark and Kathryn Yusoff, "Geosocial Formations and the Anthropocene," *Theory, Culture & Society* 34, nos. 2–3 (2017): 3–23.

47. Yusoff, "Geosocial Strata," 111.

48. Yusoff, "Geosocial Strata," 113.

49. Tina Sikka, "Geoengineering in a World Risk Society," *International Journal of Climate Change: Impacts and Responses* 3, no. 1 (2012): 149.

50. Beck, *Risk Society*, 184.

51. Beck, *Risk Society*, 186.

52. Sikka, "Geoengineering," 151.

53. Imre Szeman, "Entrepreneurship as the New Common Sense," *South Atlantic Quarterly* 114, no. 3 (July 2015): 472.

54. Michael MacCracken et al., "Alisomar Conference Statement," Asilomar International Conference on Climate Intervention Technologies, Pacific Grove, CA, March 22–26, 2010, http://climate.org.

55. MacCracken et al., "Alisomar Conference Statement."

56. Clive Hamilton, "The Clique That Is Trying to Frame the Global Geoengineering Debate," *Guardian*, December 5, 2011, https://www.theguardian.com.

57. Hamilton, *Earthmasters*, 123 (intro, n. 5).

58. Hamilton, "Clique."

59. Hamilton, *Earthmasters*, 72.

60. United Nations Office for Outer Space Affairs, Treaty on Principles Governing the Activities of States in the Exploration and Use of Outer Space, including the Moon and Other Celestial Bodies, New York, January 27, 1967, Resolution 2222 (XXI),

https://www.unoosa.org; United Nations Environment Programme, The Vienna Convention for the Protection of the Ozone Layer, Vienna, March 22, 1985, https://ozone.unep.org.

61. Royal Society, *Geoengineering the Climate*, 40; United Nations Economic Condition for Europe, Convention on Long-Range Transboundary Air Pollution, Geneva, November 13, 1979, no. 21623, https://treaties.un.org.

62. United Nations, Convention on the Prohibition of Military or any other Hostile Use of Environmental Modification Technique, New York, December 10, 1976, no. 17119, https://treaties.un.org; United Nations Environment Programme, The Montreal Protocol on Substances that Deplete the Ozone Layer, Montreal, September 15, 1987, https://www.unep.org; United Nations, Framework Convention on Climate Change, New York, May 9, 1992, no. 30822, https://treaties.un.org.

63. Intergovernmental Panel on Climate Change, *Climate Change 2014*, 89.

64. Intergovernmental Panel on Climate Change, *IPCC Expert Meeting on Geoengineering, Meeting Report*, Lima, Peru, June 20–22, 2011, https://archive.ipcc.ch, 34; see also: Jinnah Sikina, Simon Nicholson, and Jane Flegal, "Toward Legitimate Governance of Solar Geoengineering Research: A Role for Sub-State Actors," *Ethics, Policy & Environment* 21, no. 3 (2018): 362–81.

65. United Nations, Convention on Biological Diversity, "Climate-Related Geoengineering and Biodiversity: Technical and Regulatory Matters on Geoengineering in Relation to the CBD," March 23, 2017, https://www.cbd.int.

66. Paul Crutzen, "Albedo Enhancement by Stratospheric Sulfur Injections: A Contribution to Resolve a Policy Dilemma?" *Climatic Change* 77, nos. 3–4 (August 2006): 211.

67. Crutzen, "Albedo Enhancement," 212.

68. David Keith and Hadi Dowlatabadi, "A Serious Look at Geoengineering," *Eos, Transactions of the American Geophysical Union* 73, no. 27 (1992): 293.

69. Matthews and Caldeira, "Transient Climate," 9950.

70. Matthews and Caldeira, 9951.

71. Matthews and Caldeira, 9951–52.

72. Ben Kravitz et al., "An Overview of the Geoengineering Model Intercomparison Project (GeoMIP)," *Journal of Geophysical Research Atmospheres* 118, no. 23 (2013): 13105.

73. Jeff Tollefson, "First Sun-Dimming Experiment Will Test a Way to Cool Earth," *Nature*, November 27, 2018, https://www.nature.com.

74. David Keith, "Geoengineering the Climate: History and Prospect," *Annual Review of Energy and the Environment* 25 (2000): 277.

75. John Asafu-Adjaye et al., "An Ecomodernist Manifesto," https://www.ecomodernism.org, 2015.

76. Asafu-Adjaye et al., "Ecomodernist Manifesto," 30.

77. Harry Saunders, "Does Capitalism Require Endless Growth?" Breakthrough Institute, June 21, 2016, https://thebreakthrough.org.

78. Ulrich Beck, "Climate for Change, or How to Create a Green Modernity?" *Theory, Culture and Society* 27, nos. 2–3 (March 2010): 262.

79. Jason W. Moore, "Introduction: Anthropocene or Capitalocene? Nature, History, and the Crisis of Capitalism," in *Anthropocene or Capitalocene?: Nature, History, and the Crisis of Capitalism*, ed. Jason W. Moore (Oakland, CA: PM Press, 2016), 23.

80. Jason W. Moore, "The Rise of Cheap Nature," in Moore, *Anthropocene or Capitalocene?*, 240.

81. Moore, "Rise of Cheap Nature," 240.

82. International Organization for Standardization, "ISO Standard no. 14082, Working Draft 2: GHG Management—Guidance for the Quantification and Reporting of Radiative Forcing Based Climate Footprint and Mitigation Efforts," https://www.desmog.com.

83. Katie Bird, "New Draft Standard Will Help Quantify Climate Impacts of an Expanded Group of Emissions," International Organization for Standardization, August 30, 2019, https://www.iso.org.

84. Agence France-Presse, "Industry Guidance Touts Untested Tech as Climate Fix," August 23, 2019, https://www.rfi.fr.

Chapter 4

1. Mikhail I. Budyko, "The Heat Balance of the Earth's Surface," *Soviet Geography* 2, no. 4 (1961): 3–13.

2. President's Science Advisory Committee, Environmental Pollution Panel, *Restoring the Quality of Our Environment* (Washington, DC: White House, 1965,), 127, http://www.climatefiles.com.

3. Freeman J. Dyson, "Can We Control the Carbon Dioxide in the Atmosphere?" *Energy* 2, no. 3 (September 1977): 288.

4. Dyson, "Can We Control," 288.

5. Dyson, 290.

6. Teller, Wood, and Hyde, "Global Warming and Ice Ages," 3.

7. David Roochnik, *Of Art and Wisdom: Plato's Understanding of Techne* (State College: Penn State University Press, 1996), 19.

8. Roochnik, *Of Art and Wisdom*, 19.

9. Wolfgang Schadewaldt, "The Greek Concepts of 'Nature' and 'Technique,'" in *Philosophy of Technology: The Technological Condition; An Anthology*, ed. Robert C. Scharff and Val Dusek, 2nd ed. (London: Wiley Blackwell, 2014), 29.

10. Janet Atwill, *Rhetoric Reclaimed: Aristotle and the Liberal Arts Tradition* (Ithaca, NY: Cornell University Press, 1998), 2.

11. For more detail about the distinctions between techne$_1$ and techne$_2$ in a technical communication context, see: Ehren Helmut Pflugfelder, *Communicating Mobility and Technology: A Material Rhetoric for Persuasive Transportation* (London: Routledge, 2016) and idem., "Failure Matters: Conflicting Practices in a High-Tech Case." *Journal of Technical Writing and Communication* 48, no. 1 (January 2018): 31–52.

12. Atwill, *Rhetoric Reclaimed*, 85.

13. Alvin M. Weinberg, "Can Technology Replace Social Engineering?" *Bulletin of the Atomic Scientists* 22, no. 10 (1966): 4.

14. Weinberg, "Can Technology Replace" 5.

15. Weinberg, 5.

16. Evgeny Morozov, *To Save Everything, Click Here: The Folly of Technological Solutionism* (New York: Public Affairs, 2013), 6.

17. Clive Hamilton, "The Technofix Is In: A Critique of 'An Ecomodernist Manifesto,'" April 28, 2015, http://clivehamilton.com.

18. Dane Scott, "Philosophy of Technology and Geoengineering (Working Paper)," *Geoengineering Our Climate?* (blog), April 23, 2013, https://geoengineeringourclimate.wordpress.com, 159.

19. Scott, "Philosophy of Technology," 159.

20. Hamilton, *Earthmasters*, 174 (intro, n. 5).

21. Paul R. Ehrlich, *The Population Bomb* (New York: Ballantine Books, 1968). The harshest criticisms of *The Population Bomb* now hold it at least somewhat responsible for human rights abuses like large-scale sterilizations (such as in India) and/or forced abortions (such as in China). Charles C. Mann, "The Book That Incited a Worldwide Fear of Overpopulation: 'The Population Bomb' Made Dire Predictions—and Triggered a Wave of Repression Around the World," *Smithsonian Magazine*, January 2018.

22. Fred Turner, *From Counterculture to Cyberculture: Stewart Brand, the Whole Earth Network, and the Rise of Digital Utopianism* (Chicago: University of Chicago Press, 2010), 56.

23. Turner, *From Counterculture to Cyberculture*, 56.

24. Stuart Brand, ed., *Whole Earth Catalog: Access to Tools* (San Rafael, CA: Self-published, 1969).

25. Andrew G. Kirk, *Counterculture Green: The Whole Earth Catalog and American Environmentalism* (Lawrence: University Press of Kansas, 2007), 18.

26. Kirk, *Counterculture Green*, 15.

27. Norbert Wiener, *Cybernetics: Or Control and Communication in the Animal and the Machine* (Cambridge, MA: MIT Press, 1965), 39.

28. *Oxford English Dictionary Online*, s.v. "cybernetics," accessed May 17, 2020, www.oed.com.

29. Gregory Bateson, *Steps to an Ecology of Mind* (New York: Ballantine Books, 1972).

30. Bateson, *Steps to an Ecology of Mind*, 411–12.

31. Bateson, 468.

32. Turner, 124.

33. Turner, 124.

34. Bateson, 469.

35. Kirk, 175.

36. O'Neill would continue to develop his plan, with considerable backing, for years to come. He would later work for private organizations, including Brand's own Point Foundation, spend time at NASA, and take part in the Reagan-era National Commission on Space.

37. Roochnik, 53.

38. Roochnik, 52.

39. Janet Atwill, "Instituting the Art of Rhetoric: Theory, Practice, and Productive

Knowledge in Interpretations of Aristotle's Rhetoric," in *New Perspectives on the Histories of the Rhetorical Tradition*, ed. Takis Poulakos (Boulder, CO: Westview Press, 1993), 113–14.

40. Kelly Pender, *Techne: From Neoclassicism to Postmodernism: Understanding Writing as a Useful, Teachable Art* (Clemson, SC: Parlor Press, 2011), 16.

41. Atwill, *Rhetoric Reclaimed*, 2.

42. Atwill, *Rhetoric Reclaimed*, 85; 84.

43. John Wild, "Plato's Theory of Texnh: A Phenomenological Interpretation," *Philosophy and Phenomenological Research* 1, no. 3 (March 1941): 257.

44. Rika Preiser, Laura M. Pereira, and Reinette (Oonise) Biggs, "Navigating Alternative Framings of Human-Environment Interactions: Variations on the Theme of 'Finding Nemo,'" *Anthropocene* 20 (December 2017): 84.

45. Ted Nordhaus and Michael Shellenberger, "The Death of Environmentalism: Global Warming Politics in a Post-Environmental World," Breakthrough Institute, September 24, 2004, https://thebreakthrough.org, 6.

46. Nordhaus and Shellenberger, "Death of Environmentalism," 12.

47. Asafu-Adjaye et al., "Ecomodernist Manifesto," 6.

48. Asafu-Adjaye et al., 21.

49. International Maritime Organization, Convention on the Prevention of Marine Pollution by Dumping Wastes and Other Matter, London, December 29, 1972, https://www.imo.org; United Nations, Convention on Biological Diversity, Rio de Janeiro, June 5, 1992, no. 30619, https://www.cbd.int; Parliament of Canada, Canadian Environmental Protection Act, 1999, SC 1999, c 33, https://canlii.ca; United Nations, Cartagena Protocol on Biosafety to the Convention on Biological Diversity, Cartagena, May 16, 2000, no. 30619, https://bch.cbd.int; Martin Lukacs, "World's Biggest Geoengineering Experiment 'Violates' UN Rules," *Guardian*, October 15, 2012, https://www.theguardian.com.

50. John Weier, "John Martin (1935–1993)," Earth Observatory (NASA), July 10, 2001, https://earthobservatory.nasa.gov.

51. John H. Martin and Steve E. Fitzwater, "Iron Deficiency Limits Phytoplankton Growth in the North-East Pacific Subarctic," *Nature* 331, no. 6154 (1988): 343.

52. John H. Martin, Steve E. Fitzwater, and R. Michael Gordon, "Iron Deficiency Limits Phytoplankton Growth in Antarctic Waters," *Global Biogeochemical Cycles* 4, no. 1 (March 1990): 5–12.

53. Martin, Fitzwater, and Gordon, "Iron Deficiency," 5.

54. Kenneth H. Coale et al., "A Massive Phytoplankton Bloom Induced by an Ecosystem-Scale Iron Fertilization Experiment in the Equatorial Pacific Ocean," *Nature* 383, no. 6600 (1996): 500.

55. Aaron L. Strong, John J. Cullen, and Sallie W. Chisholm. "Ocean Fertilization: Science, Policy, and Commerce," *Oceanography* 22, no. 3 (2009): 239.

56. Coale et al., "Massive Phytoplankton Bloom," 500.

57. Coale et al., 501.

58. Strong, Cullen, and Chisholm, "Ocean Fertilization," 239–40.

59. Strong, Cullen, and Chisholm, 240.

60. This list of tests includes: SOIREE (1999), EisenEx (2000), SEEDS-I (2001),

SOFeX-N (2002), SOFeX-S (2001), SERIES (2002), EIFEX (2004), FeeP (2004), SAGE (2004), and SEEDS-II (2004).

61. Ken O. Buesseler and Philip W. Boyd, "Will Ocean Fertilization Work?" *Science* 300, no. 5616 (2003): 67–68; Hein J. W. De Baar et al., "Synthesis of Iron Fertilization Experiments: From the Iron Age in the Age of Enlightenment," *Journal of Geophysical Research: Oceans* 110, no. C9 (2005): 1–24.

62. Sallie Chisholm, Paul G. Falkowski, and John J. Cullen, "Dis-Crediting Ocean Fertilization," *Science* 294, no. 5541 (2001): 310.

63. Oliver W. Wingenter et al., "Changing Concentrations of CO, CH_4, C_5H_8, CH_3Br, CH_3I, and Dimethyl Sulfide during the Southern Ocean Iron Enrichments Experiments," *PNAS* 101, no. 23 (2004): 8540; Patricia M. Gilbert et al., "Ocean Urea Fertilization for Carbon Credits Poses High Ecological Risks," *Marine Pollution Bulletin* 56, no. 6 (June 2008): 1054.

64. John Cullen and Philip Boyd, "Predicting and Verifying the Intended and Unintended Consequences of Large-Scale Ocean Iron Fertilization," *Marine Ecology Progress Series* 364 (2008): 295–301.

65. Strong, Cullen, and Chisholm, 247.

66. Strong, Cullen, and Chisholm, 249.

67. Charles Graeber, "Dumping Iron," *Wired*, November 1, 2000, https://www.wired.com.

68. Ocean Nourishment Corporation, "Agenda: Waste Atmospheric Carbon Dioxide," accessed April 14, 2020, www.oceannourishment.com.

69. Sonia D. Batten and James F. R. Gower, "Did the Iron Fertilization near Haida Gwaii in 2012 Affect the Pelagic Lower Trophic Level Ecosystem?" *Journal of Plankton Research* 36, no. 4 (July–August 2014): 925–32.

70. Luokkanen, Huttunen, and Hildén, "Geoengineering, News Media and Metaphors," 978.

71. Graeber, "Dumping Iron."

72. David Beillo, *The Unnatural World: The Race to Remake Civilization in Earth's Newest Age* (New York: Scribner, 2016), 35; Levitt and Dubner, *SuperFreakonomics*, 184.

73. Haida Salmon Restoration Corporation, "Ocean Micro Nutrient Replenishment," November 2, 2018, https://www.haidasalmonrestoration.com.

74. "Iron Fertilization," Living on Earth, https://loe.org.

75. Haida Salmon Restoration Corporation, "Ocean Micro Nutrient Replenishment."

76. ETC Group, "Informational Backgrounder on the 2012 Haida Gwaii Iron Dump," March 3, 2013, https://www.etcgroup.org.

77. Kate Elizabeth Gannon and Mike Hulme, "Geoengineering at the 'Edge of the World': Exploring Perceptions of Ocean Fertilisation through the Haida Salmon Restoration Corporation," *Geo: Geography and Environment* 5, no. 1 (January–June 2018): 3.

78. They should have also been aware of the 2009 Anchorage Declaration, which called geoengineering a "false solution." The declaration states, "when specific programs and projects affect our lands, territories, environment and natural resources, the right of Self Determination of Indigenous Peoples must be recognized and respected,

emphasizing our right to Free, Prior and Informed Consent, including the right to say "no"." See: Indigenous Peoples' Global Summit on Climate Change, "The Anchorage Declaration," Anchorage, April 24, 2009, https://unfccc.int, 3.

79. Joseph Weiss, *Shaping the Future on Haida Gwaii: Life beyond the Settler Colonialism* (Victoria: University of British Columbia Press, 2018), 50.

80. Gannon and Hulme, "Geoengineering," 3; Yusoff, "Anthropogenesis," 5.

81. Zach Horton, "Going Rogue or Becoming Salmon?: Geoengineering Narratives in Haida Gwaii," *Cultural Critique* 97 (2017), 132.

82. Haida Salmon Restoration Corporation, "Our Mission," accessed March 2, 2020, http://www.haidasalmonrestoration.com/.

83. Lukacs, "World's Biggest."

84. ETC Group, "Informational Backgrounder."

85. Gannon and Hulme, 4.

86. Gannon and Hulme, 4.

87. Horton, "Going Rogue" 146.

88. Gannon and Hulme, 5.

89. Horton, 154.

90. Horton, 141.

91. Batten and Gower, "Did the Iron Fertilization," 931.

92. ETC Group.

93. Gannon and Hulme, 4.

94. Naomi Klein, "Geoengineering: Testing the Waters," *New York Times*, October 28, 2012, https://www.nytimes.com.

95. Holly Jean Buck, *After Geoengineering: Climate Tragedy, Repair, and Restoration* (London: Verso, 2019), 167.

96. Gabriel Lopez et al., "Carbon Removal Technologies," *Y Combinator*, https://www.ycombinator.com.

97. The phrase "move fast and break things" has been ascribed to Facebook founder Mark Zuckerberg.

Chapter 5

1. Paul J. Crutzen and Eugene F. Stoermer, "The 'Anthropocene,'" *Global Change Newsletter* 41, no. 1 (May 2000), 17.

2. Crutzen and Stoermer, "The 'Anthropocene,'" 18.

3. Crutzen and Stoermer, 18.

4. For some examples of these references to the noosphere, see: Crutzen, "Geology of Mankind," 23; Jan Zalasiewicz et al., "The New World of the Anthropocene," *Environmental Science and Technology* 44, no. 7 (2010): 2228–31; Will Steffen et al., "The Anthropocene: Conceptual and Historical Perspectives," *Philosophical Transactions of the Royal Society* 369, no. 1938 (2011): 842–67.

5. W. Lucht and R. K. Pachauri, "The Mental Component of the Earth System," in *Earth System Analysis for Sustainability*, ed. Hans Joachim Schellnhuber et al. (Cambridge, MA: MIT Press, 2004), 343.

6. Yusoff, *Billion Black Anthropocenes*, 86.

7. Eduard Suess, *The Face of the Earth / Das Antlitz der Erde* (Oxford: Clarendon Press, 1904).

8. Vladimir I. Vernadsky, "Geochemistry," in *The Biosphere and Noosphere Reader: Global Environment, Society, and Change*, ed. Paul R. Samson and David Pitt (London: Routledge, 2012), 26–27; Vladimir I. Vernadsky, "The Biosphere," in *The Biosphere and Noosphere Reader: Global Environment, Society, and Change*, ed. Paul R. Samson and David Pitt (London: Routledge, 2012), 35–38.

9. Vernadsky, "Geochemistry," 27.

10. Bertrand Guillaume, "Vernadsky's Philosophical Legacy: A Perspective from the Anthropocene," *Anthropocene Review* 1, no. 2 (August 2014): 141.

11. Clive Hamilton and Jacques Grinevald, "Was the Anthropocene Anticipated?" *Anthropocene Review* 2, no. 1 (April 2015): 66.

12. Matthew Lepori, "There Is No Anthropocene: Climate Change, Species-Talk, and Political Economy," *Telos* 172 (Fall 2015): 110.

13. Chakrabarty, "Climate of History."

14. Malm and Hornborg, "Geology of Mankind?" 65.

15. David Satterthwaite, "The Implications of Population Growth and Urbanization for Climate Change," *Environment and Urbanization* 21, no. 2 (October 2009): 564.

16. Bonneuil, "Geological Turn," 19.

17. Bonneuil, 23.

18. Andrew C. Revkin, "Managing Planet Earth; Forget Nature. Even Eden Is Engineered," *New York Times*, August 20, 2002, https://www.nytimes.com.

19. Vladimir I. Vernadsky, "The Biosphere," in Samson and Pitt, *Biosphere and Noosphere Reader*, 8–10.

20. Crutzen and Stoermer, 18.

21. Crutzen, "Geology of Mankind," 23.

22. Bonneuil, 23.

23. *World Heritage Encyclopedia*, s.v. "homo faber," www.gutenberg.us.

24. Vernadsky, "Geochemistry," 27.

25. Bergson, *Creative Evolution*, 153. Emphasis in original.

26. Daniel Chernilo, "The Question of the Human in the Anthropocene Debate," in "The Anthropocene and Social Theory 20th Anniversary Special Issue," special issue, *European Journal of Social Theory* 20, no. 1 (February 2017): 54.

27. Chernilo, "Question of the Human," 56.

28. Bonneuil, 23.

29. Chernilo, 54.

30. Crutzen, "Geology of Mankind," 23.

31. Hamilton, "Technofix Is In."

32. Mark Lynas, *The God Species: Saving the Planet in the Age of Humans* (Washington, DC: National Geographic, 2011).

33. Lynas, *God Species*, 10.

34. Lynas, 6.

35. Intergovernmental Panel on Climate Change, *Climate Change 2014*, 1108.

36. Albert Lin's work details how these psychological heuristics impact human

ability to act in sustainable, ecologically sound ways; it also considers the issue of a moral hazard. Albert C. Lin, "Does Geoengineering Present a Moral Hazard?" *Ecology Law Quarterly* 40, no. 3 (2013): 673–712.

37. Daniel Kahneman, *Thinking, Fast and Slow* (New York: Farrar, Straus and Giroux, 2011).

38. George Marshall, *Don't Even Think About It: Why Our Brains Are Wired to Ignore Climate Change* (New York: Bloomsbury, 2014), 58.

39. Stormer and McGreavy, "Thinking Ecologically," 5.

40. Stormer and McGreavy, 5.

41. Stormer and McGreavy, 5.

42. Stormer and McGreavy, 5.

43. Kenneth Burke, *Language as Symbolic Action: Essays on Life, Literature, and Method* (Berkeley: University of California Press, 1966), 16.

44. Stormer and McGreavy, 6.

45. Stormer and McGreavy, 18.

46. Stormer and McGreavy, 18.

47. Stormer and McGreavy, 19.

48. Stormer and McGreavy, 18.

49. Elizabeth A. Flynn, Patricia Sotirin, and Ann Brady, eds., *Feminist Rhetorical Resilience* (Logan: Utah State University Press, 2012), 7.

50. R. T. Paine, "A Note on Trophic Complexity and Community Stability," *American Naturalist* 103, no. 929 (1969): 91–93.

51. C. Folke et al., "Social-Ecological Resilience and Biosphere-Based Sustainability Science," *Ecology and Society* 21, no. 3 (September 2016): 9.

52. C. Folke et al., "Social-Ecological Resilience," 9.

53. Erle C. Ellis, *Anthropocene: A Very Short Introduction* (Oxford: Oxford University Press, 2018), 151.

54. Erle C. Ellis, "The Planet of No Return: Human Resilience on an Artificial Earth," *Breakthrough Journal* 2 (Fall 2011): 38.

55. Folke et al., 9.

56. C. S. Holling, "Resilience and Stability of Ecological Systems," *Annual Review of Ecology and Systematics* 4 (1973): 14. Emphasis mine.

57. C. S. Holling, "Engineering Resilience versus Ecological Resilience," in *Engineering within Ecological Constraints*, ed. Peter C. Schulze (Washington, DC: National Academies Press, 1996), 33.

58. Holling, "Resilience and Stability," 17.

59. Holling, "Engineering Resilience," 33.

60. Holling, "Engineering Resilience," 33.

61. Intergovernmental Panel on Climate Change, *Special Report: Global Warming of 1.5°C.*

62. *Disruption*, directed by Kelly Nyks and Jared P. Scott (PF Pictures, 2014).

63. Brian Walsh et al., "Pathways for Balancing CO_2 Emissions and Sinks," *Nature Communications* 8, no. 14856 (2017), 1.

64. Chen and Massimo Tavoni, "Direct Air Capture of CO_2 and Climate

Stabilization: A Model Based Assessment," *Climatic Change* 118, no. 1 (May 2013): 63. Emphasis mine.

Chapter 6

1. Jean-Francois Bastin et al., "The Global Tree Restoration Potential," *Science* 365, no. 6448 (2019): 78.

2. Buck, *After Geoengineering*, 116. Emphasis in original.

3. Matt McGrath, "Key 'Step Forward' in Cutting Cost of Removing CO_2 from Air," *BBC News*, June 7, 2018, https://www.bbc.com.

4. Robert Socolow et al., *Direct Air Capture of CO2 with Chemicals*, report presented for the American Physical Society Panel on Public Affairs, June 1, 2011, https://www.aps.org.

5. Environmental Defense Fund, "Our Position on Geoengineering," accessed June 3, 2020, https://www.edf.org; "Hands Off Mother Earth! Manifesto against Geoengineering," *Geoengineering Monitor*, October 4, 2018, https://www.geoengineeringmonitor.org.

6. Greenpeace, USA, *Carbon Capture Scam (CCS): How a False Climate Solution Bolsters Big Oil* (Washington, DC: Greenpeace USA), 2015, 5.

7. Nixon, *Slow Violence*, 2.

8. David Keith, "A Critical Look at Geoengineering against Climate Change," *TEDSalon*, 2007, https://www.ted.com; Martin Bunzl, "Researching Geoengineering: Should Not or Could Not?" *Environmental Research Letters* 4, no. 4 (December 2009): 45104.

9. Jesse Reynolds, "A Critical Examination of the Climate Engineering Moral Hazard and Risk Compensation Concern," *Anthropocene Review* 2, no. 2 (August 2015): 177.

10. Andrew Lockley and D'Maris Coffman, "Distinguishing Morale Hazard from Moral Hazard in Geoengineering," *Environmental Law Review* 18, no. 3 (2016): 194–204.

11. Keith, "Geoengineering the Climate," 276.

12. Royal Society, *Geoengineering the Climate*, 4; 37.

13. United States Government Accountability Office, *Climate Engineering: Technical Status, Future Directions, and Potential Responses*, Report to Congressional Requester, GAO-11-71, July 2011, https://www.gao.gov, 67.

14. Marion Hourdequin, "Geoengineering, Solidarity, and Moral Risk," in *Engineering the Climate: The Ethics of Solar Radiation Management*, ed. Christopher J. Preston (Lanham, MD: Lexington Books, 2012), 26.

15. Corner and Pidgeon, "Geoengineering, Climate Change Scepticism," 12.

16. Victoria Wibeck et al., "Making Sense of Climate Engineering: A Focus Group Study of Lay Publics in Four Countries," *Climatic Change* 145, nos. 1–2 (November 2017): 8.

17. Christine Merk, Gert Pönitzsch, and Katrin Rehdanz, "Knowledge about Aerosol Injection Does Not Reduce Individual Mitigation Efforts," *Environmental Research Letters* 11, no. 5 (2016): 1–6.

18. Dan M. Kahan et al., "Geoengineering and Climate Change Polarization: Testing a Two-Channel Model of Science Communication," *Annals of the American Academy of Political and Social Science* 658 (2015): 192–222; Malcolm Fairbrother,

"Geoengineering, Moral Hazard, and Trust in Climate Science: Evidence from a Survey Experiment in Britain," *Climatic Change* 139, nos. 3–4 (December 2016): 480.

19. Fairbrother, "Geoengineering, Moral Hazard," 487.

20. Kahan et al., "Geoengineering and Climate Change Polarization," 192. Emphasis mine.

21. Kahan et al., 204.

22. For these definitions of a moral hazard, see: Edward A. Parson, "Reflections on Air Capture: The Political Economy of Active Intervention in the Global Environment," *Climatic Change* 74, no. 1 (2006): 5–15; National Research Council, *Advancing the Science of Climate Change* (Washington, DC: National Academies Press, 2010); Lin, "Does Geoengineering Present a Moral Hazard?"; Hamilton, *Earthmasters*. For their part, Kahan et al. seem to employ, but also disregard, these definitions.

23. David W. Keith, Edward Parson, and M. Granger Morgan, "Research on Global Sun Block Needed Now," *Nature* 463, no. 7280 (2010): 427.

24. For more detail about social media's influence on global collective behavior and the environment, see Joseph B. Bak-Coleman et al., "Stewardship of Global Collective Behavior., *Proceedings of the National Academy of Sciences* 118, no. 27 (2021), e2025764118.

25. United Nations Environment Programme, *Decoupling 2: Technologies, Opportunities and Policy Options. A Report of the Working Group on Decoupling to the International Resource Panel*, 2014, https://www.unep.org, 2.

26. Asafu-Adjaye et al., "Ecomodernist Manifesto," 25, 11.

27. Asafu-Adjaye et al., 28.

28. Marion Hourdequin, "Climate Change, Climate Engineering, and the 'Global Poor': What Does Justice Require?" *Ethics, Policy & Environment* 21, no. 3 (2018): 277.

29. Yusoff, *Billion Black Anthropocenes*, 14.

30. Yusoff, *Billion*, 19.

31. Yusoff, *Billion*, 16.

32. David Keith and Joshua Horton, "Solar Geoengineering and Obligations to the Global Poor," in *Climate Justice and Geoengineering: Ethics and Policy in the Atmospheric Anthropocene*, ed. Christopher J. Preston (London: Rowman and Littlefield, 2016), 87.

33. Keith and Horton, "Solar Geoengineering and Obligations," 90.

34. Nordhaus and Shellenberger, "Death of Environmentalism," 10.

35. Yusoff, *Billion*, 15.

36. Wylie Carr and Christopher J. Preston, "Skewed Vulnerabilities and Moral Corruption in Global Perspectives on Climate Engineering," *Environmental Values* 26, no. 6 (December 2017): 766.

37. For a companion discussion to Carr and Preston's work on developing countries and geoengineering, see Kyle Powys Whyte's (2018) commentary on Indigenous perspectives on geoengineering. Kyle Powys White, "Indigeneity in Geoengineering Discourses: Some Considerations," *Ethics, Policy & Environment* 21, vol. 3 (2018): 289–307.

38. Yusoff, *Billion*, 37.

39. Yusoff, *Billion*, 22.

40. Yusoff, *Billion*, 22.

41. Yusoff, *Billion*, 36.

42. Nixon, 3.

43. Eloy S. Sanz Sanz-Pérez et al., "Direct Capture of CO_2 from Ambient Air," *Chemical Reviews* 116, no. 19 (2016).

44. Yusoff, *Billion*, 36.

Conclusion

1. Extinction Rebellion, accessed June 24, 2020, https://extinctionrebellion.us.

2. Extinction Rebellion America, "Who We Are," accessed June 24, 2020, https://xramerica.org.

3. Geoff Dembicki, "A Debate Over Racism Has Spread to One of the World's Most Famous Climate Groups," *Vice*, April 28, 2020, https://www.vice.com.

4. Mike Hulme, *Why We Disagree about Climate Change* (Cambridge, UK: Cambridge University Press, 2009), v.

5. M. D. King et al., "Dynamic Ice Loss from the Greenland Ice Sheet Driven by Sustained Glacier Retreat," *Communications Earth Environment* 1, no. 1 (2020).

6. Abi Bradford, Sylvia Broude, and Alexander Truelove, "Trash in America: Moving from a Destructive Consumption to a Zero-Waste System," Frontier Group, February 12, 2018, https://frontiergroup.org; *Frontline*, episode 8, "Plastic Wars," produced by Rick Young et al., aired March 31, 2020, on PBS; International Energy Agency, "Global Energy Demand to Plunge This Year as a Result of the Biggest Shock Since the Second World War," April 30, 2020, https://www.iea.org.

7. Geoff Mann and Joel Wainwright, *Climate Leviathan: A Political Theory of Our Planetary Future* (London: Verso, 2018), 88.

8. Mann and Wainwright, *Climate Leviathan*, 90.

9. John Urry, *Mobilities* (London: Polity, 2007), 289.

10. Naomi Klein, *This Changes Everything: Capitalism vs. the Climate* (New York: Simon and Schuster, 2014), 294–95.

11. Klein, *This Changes Everything*, 295.

12. Christina Henriksen et al., Letter to the SCoPEx Advisory Committee, February 24, 2021.

13. Henriksen et al, Letter.

14. Henriksen et al.

15. SCoPEx Advisory Committee, "The SCoPEx Advisory Committee Is Updating and Diversifying Its Membership," https://scopexac.com.

16. Aristotle, "Metaphysics," in Barnes, *Complete Works of Aristotle*, 1022b4–1022b14.

17. Nixon, *Slow Violence*, 238.

18. Climate Justice Alliance, "What Do We Mean by a Just Transition?" May 16, 2018, https://climatejusticealliance.org.

19. Yusoff, *Billion Black Anthropocenes*, 109.

BIBLIOGRAPHY

Agence France-Presse. "Industry Guidance Touts Untested Tech as Climate Fix." August 23, 2019. https://www.rfi.fr/en/.

Anshelm, Jonas, and Anders Hansson. "'The Last Chance to Save the Planet?: An Analysis of the Geoengineering Advocacy Discourse in the Public Debate." *Environmental Humanities* 5, no. 1 (May 2014): 101–23.

Arènes, Alexandra, Jérôme Gaillardet, and Bruno Latour. "Giving Depth to the Surface: An Exercise in the Gaia-Graphy of Critical Zones." *Anthropocene Review* 5, no. 2 (August 2018): 120–35.

Aristotle. *The Complete Works of Aristotle*. Translated and edited by Jonathan Barnes. Princeton: Princeton University Press, 1984.

Arrhenius, Svante. "On the Influence of Carbonic Acid in the Air upon the Temperature of the Ground." *London, Edinburgh, and Dublin Philosophical Magazine and Journal of Science* 41, no. 251 (1896): 237–76.

Asafu-Adjaye, John, Linus Blomqvist, Stewart Brand, et al. "An Ecomodernist Manifesto." www.ecomodernism.org. 2015.

Atwill, Janet. "Instituting the Art of Rhetoric: Theory, Practice, and Productive Knowledge in Interpretations of Aristotle's Rhetoric." In *New Perspectives on the Histories of the Rhetorical Tradition*, edited by Takis Poulakos, 91–117. Boulder, CO: Westview Press, 1993.

———. *Rhetoric Reclaimed: Aristotle and the Liberal Arts Tradition*. Ithaca, NY: Cornell University Press, 1998.

Baake, Ken. *Metaphor and Knowledge: The Challenges of Writing Science*. Albany: SUNY Press, 2003.

Badiou, Alain. *Metapolitics*. London: Verso, 2005.

Bak-Coleman, Joseph B., Mark Alfano, Wolfram Barfuss, et al. "Stewardship of Global Collective Behavior." *Proceedings of the National Academy of Sciences* 118, no. 27 (2021): e2025764118.

Barad, Karen Michelle. *Meeting the Universe Halfway: Quantum Physics and the Entanglement of Matter and Meaning*. Durham, NC: Duke University Press, 2007.

Barnes, Barry. *Scientific Knowledge and Sociological Theory*. London: Routledge and K. Paul, 1974.

Barnes, Jonathan. *The Presocratic Philosophers*. London: Routledge, 1982.

Barnett, Scot, and Casey Boyle, eds. *Rhetoric, through Everyday Things*. Tuscaloosa: University of Alabama Press, 2016.

Baskin, Morgan, and Jack Denton. "'The Ultimate Patent Troll': In the Words of

Technology Reporters." Pacific Standard, August 23, 2017; updated September 16, 2018, https://psmag.com.

Bastin, Jean-Francois, Yelena Finegold, Claude Garcia, et al., "The Global Tree Restoration Potential." *Science* 365, no. 6448 (2019): 76–79.

Bateson, Gregory. *Steps to an Ecology of Mind*. New York: Ballantine Books, 1972.

Batten, Sonia D., and James F. R. Gower. "Did the Iron Fertilization near Haida Gwaii in 2012 Affect the Pelagic Lower Trophic Level Ecosystem?" *Journal of Plankton Research* 36, no. 4 (July–August 2014): 925–32.

Beck, Ulrich. "Climate for Change, or How to Create a Green Modernity?" *Theory, Culture and Society* 27, nos. 2–3 (March 2010): 254–66.

———. "Incalculable Futures: World Risk Society and Its Social and Political Implications." In *Ulrich Beck: Pioneer in Cosmopolitan Sociology and Risk Society*, edited by Ulrich Beck, 78–89. Heidelberg: Springer, 2014.

———. "The Reinvention of Politics: Towards a Theory of Reflexive Modernization." In *Reflexive Modernization: Politics, Tradition and Aesthetics in the Modern Social Order*, edited by Ulrich Beck, Anthony Giddens, and Scott Lash, 1–55. Stanford, CA: Stanford University Press, 1994.

———. *Risk Society: Towards a New Modernity*. Translated by Mark Ritter. London: SAGE, 1992.

Beillo, David. *The Unnatural World: The Race to Remake Civilization in Earth's Newest Age*. New York: Scribner, 2016.

Bellamy, Rob, Jason Chilvers, Naomi E. Vaughan, and Timothy M. Lenton. "A Review of Climate Geoengineering Appraisals." *Wiley Interdisciplinary Reviews: Climate Change* 3, no. 6 (November–December 2012): 597–615.

Bergson, Henri. *Creative Evolution*. Translated by Arthur Mitchell. New York: Henry Holt, 1911.

Berman, Russel. "The Senate Officially Believes Climate Change Is Real." *Atlantic*, January 21, 2015. https://www.theatlantic.com.

Bickel, J. Eric, and Lee Lane. "An Analysis of Climate Engineering as a Response to Climate Change." Copenhagen Consensus Center, August 7, 2009. https://copenhagenconsensus.com.

Biggs, Andy. "Congressman Andy Biggs' Opening Statement on Joint Subcommittee Hearing on Geoengineering." November 8, 2017. https://biggs.house.gov.

Bird, Katie. "New Draft Standard Will Help Quantify Climate Impacts of an Expanded Group of Emissions." International Organization for Standardization, August 30, 2019. https://www.iso.org.

Bloor, David. *Knowledge and Social Imagery*. London: Routledge, 1976.

Boettcher, Miranda, and Stefan Schäfer. "Reflecting upon 10 Years of Geoengineering Research: Introduction to the Crutzen + 10 Special Issue." *Earth's Future* 5, no. 3 (March 2017): 266–77.

Bonneuil, Cristophe. "The Geological Turn: Narratives of the Anthropocene." In *The Anthropocene and the Global Environmental Crisis*, edited by Clive Hamilton, François Gemenne, and Christophe Bonneuil, 17–31. London: Routledge, 2015.

Bonneuil, Christophe, and Jean-Baptiste Fressoz. *The Shock of the Anthropocene: The Earth, History and Us*. Translated by David Fernbach. London: Verso, 2017.

Bonta, Mark, and John Protevi. *Deleuze and Geophilosophy: A Guide and Glossary*. Edinburgh: Edinburgh University Press, 2004.

Boulding, Kenneth E. "The Economics of the Coming Spaceship Earth." In *Environmental Quality in a Growing Economy: Essays from the Sixth RFF Forum*, edited by H. Jarrett, 3–14. Baltimore: Johns Hopkins University Press, 1966.

Boyd, Philip W. "Ranking Geo-Engineering Schemes." *Nature Geoscience* 1, no. 11 (2008): 722–24.

Bradford, Abi, Sylvia Broude, and Alexander Truelove. "Trash in America: Moving from a Destructive Consumption to a Zero-Waste System." Frontier Group, February 12, 2018. https://frontiergroup.org.

Brand, Stewart, ed. *Whole Earth Catalog: Access to Tools*. San Rafael, CA: Self-published, 1969.

Brown, Jim, and Nathaniel Rivers. "Composing the Carpenter's Workshop." *Journal of Object Oriented Studies* 1 (2013): 27–36.

Brulle, Robert J. "Institutionalizing Delay: Foundation Funding and the Creation of Climate Change Counter-Movement Organizations." *Climatic Change* 122, no. 4 (February 2014): 681–94.

Budyko, Mikhail I. "The Heat Balance of the Earth's Surface." *Soviet Geography* 2, no. 4 (1961): 3–13.

Buck, Holly Jean. *After Geoengineering: Climate Tragedy, Repair, and Restoration*. London: Verso, 2019.

Buesseler, Ken O., and Philip W. Boyd. "Will Ocean Fertilization Work?" *Science* 300, no. 5616 (2003): 67–68.

Bunzl, Martin. "Researching Geoengineering: Should Not or Could Not?" *Environmental Research Letters* 4, no. 4 (December 2009): 45104.

Burke, Kenneth. *Language as Symbolic Action: Essays on Life, Literature, and Method*. Berkeley: University of California Press, 1966.

Callendar, G. S. "The Artificial Production of Carbon Dioxide and Its Influence on Temperature." *Quarterly Journal of the Royal Meteorological Society* 64, no. 275 (1938): 223–40.

Calma, Justine. "As a Last Resort, Andrew Yang Proposes Space Mirrors to Save the Planet." Verge, August 26, 2019. https://www.theverge.com.

Carr, Wylie, and Christopher J. Preston. "Skewed Vulnerabilities and Moral Corruption in Global Perspectives on Climate Engineering." *Environmental Values* 26, no. 6 (December 2017): 757–77.

Carson, Rachel. *Silent Spring*. Boston: Houghton Mifflin, 1962.

Carter, Paul. *The Lie of the Land*. London: Faber and Faber, 1996.

Castree, Noel, and Thomas MacMillan. "Dissolving Dualisms: Actor-Networks and the Reimagination of Nature." In *Social Nature: Theory, Practice, and Politics*, edited by Noel Castree and Bruce Braun, 208–24. Oxford: Blackwell, 2001.

Ceccarelli, Leah. "Manufactured Scientific Controversy: Science, Rhetoric, and Public Debate." *Rhetoric and Public Affairs* 14, no. 2 (Summer 2011): 195–228.

Chakrabarty, Dipesh. "The Climate of History: Four Theses." *Critical Inquiry* 35, no. 2 (Winter 2009): 197–222.

Chen, Mel Y. *Animacies: Biopolitics, Racial Mattering, and Queer Affect*. Durham, NC: Duke University Press, 2012.

Chen, Chen, and Massimo Tavoni. "Direct Air Capture of CO_2 and Climate Stabilization: A Model Based Assessment." *Climatic Change* 118, no. 1 (May 2013): 59–72.

Chernilo, Daniel. "The Question of the Human in the Anthropocene Debate." In "The Anthropocene and Social Theory 20th Anniversary Special Issue." Special issue, *European Journal of Social Theory* 20, no. 1 (February 2017): 44–60.

Chisholm, Sallie, Paul G. Falkowski, and John J. Cullen. "Dis-Crediting Ocean Fertilization." *Science* 294, no. 5541 (2001): 309–10.

Crichton, Jonathon, Christopher N. Candlin, and Arthur S. Firkins, eds. *Communicating Risk: Communicating in Professions and Organizations*. London: Palgrave Macmillan, 2016.

Chris, Robert. *Systems Thinking for Geoengineering Policy: How to Reduce the Threat of Dangerous Climate Change by Embracing Uncertainty and Failure*. London: Routledge, 2015.

Cicero. *Topica*. Translated by Tobias Reinhardt. Oxford: Oxford University Press, 2003.

Clark, Nigel. "Geoengineering and Geologic Politics." *Environment and Planning A* 45, no. 12 (December 2013): 2825–32.

———. *Inhuman Nature: Sociable Life on a Dynamic Planet*. London: SAGE, 2011.

———. "Politics of Strata." *Theory, Culture and Society* 34, nos. 2–3 (March–May 2017): 211–31.

Clark, Nigel, and Kathryn Yusoff. "Geosocial Formations and the Anthropocene." *Theory, Culture and Society* 34, no. 2–3 (2017): 3–23.

Clary-Lemon, Jennifer. *Planting the Anthropocene: Rhetorics of Natureculture*. Logan: Utah State University Press, 2019.

Climate Justice Alliance. "What Do We Mean by a Just Transition?" May 16, 2018. https://climatejusticealliance.org.

Coale, Kenneth H., Kenneth S. Johnson, Steve E. Fitzwater, et al. "A Massive Phytoplankton Bloom Induced by an Ecosystem-Scale Iron Fertilization Experiment in the Equatorial Pacific Ocean." *Nature* 383, no. 6600 (1996): 495–501.

Connolly, William E. *Climate Machines, Fascist Drives, and Truth*. Durham, NC: Duke University Press, 2019.

Consigny, Scott Porter. *Gorgias, Sophist and Artist*. Columbia: University of South Carolina Press, 2001.

Coole, Diana. "The Inertia of Matter and the Generativity of Flesh." In *New Materialisms: Ontology, Agency, and Politics*, edited by Diana Coole and Samantha Frost, 92–115. Durham, NC: Duke University Press, 2010.

Corner, Adam, and Nick Pidgeon. "Geoengineering, Climate Change Scepticism and the 'Moral Hazard' Argument: An Experimental Study of UK Public Perceptions." *Philosophical Transactions. Series A* 372, no. 2031 (2014): 1–14.

———. "Like Artificial Trees? The Effect of Framing by Natural Analogy on Public Perceptions of Geoengineering." *Climatic Change* 130, no. 3 (June 2015): 425–38.

Cosmos: A Personal Voyage. Episode 1, "The Shores of the Cosmic Ocean." Directed by Adrian Malone, Geoffrey Haines-Stiles, Rob McCain, David F. Oyster, Tom Weidlinger, and Richard J. Wells, written by Carl Sagan, Ann Druyan, and Steven Soter, hosted by Carl Sagan, aired September 28, 1980, on PBS.

Cressey, Daniel. "Latest IPCC Climate Report Puts Geoengineering in the Spotlight." *Scientific American*, October 2, 2013. https://www.scientificamerican.com.

Crutzen, Paul. "Albedo Enhancement by Stratospheric Sulfur Injections: A Contribution to Resolve a Policy Dilemma?" *Climatic Change* 77, no. 3–4 (August 2006): 211–20.

———. "Geology of Mankind." *Nature* 415, no. 6867 (2002): 23.

Crutzen, Paul J., and Eugene F. Stoermer. "The 'Anthropocene.'" *Global Change Newsletter* 41, no. 1 (May 2000): 17–18.

Cullen, John, and Philip Boyd. "Predicting and Verifying the Intended and Unintended Consequences of Large-Scale Ocean Iron Fertilization." *Marine Ecology Progress Series* 364 (2008): 295–301.

Danisch, Robert. "Political Rhetoric in a World Risk Society." *Rhetoric Society Quarterly* 40, no. 2 (2010): 172–92.

Datson, Loraine J., and Peter Galison. *Objectivity*. Cambridge, MA: MIT Press, 2010.

De Baar, Hein J. W., Philip W. Boyd, Kenneth H. Coale, et al. "Synthesis of Iron Fertilization Experiments: From the Iron Age in the Age of Enlightenment." *Journal of Geophysical Research: Oceans* 110, no. C9 (2005): 1–24.

Debeir, Jean-Paul, Deléage, Jean-Claude, and Daniel Hémery. *In the Servitude of Power: Energy and Civilization through the Ages*. London: Zed Books, 1991.

DeLanda, Manuel. "Deleuzian Social Ontology and Assemblage Theory." In *Deleuze and the Social*, edited by Martin Fuglsang and Bent Meier Sorensen, 250–66. Edinburgh: Edinburgh University Press, 2006.

Deleuze, Gilles, and Félix Guattari. *A Thousand Plateaus: Capitalism and Schizophrenia*. Translation and forward by Brian Massumi. Minneapolis: University of Minnesota Press, 1987.

Deleuze, Gilles, and Claire Parnet. *Dialogues*. Translated by Hugh Tomlinson and Barbara Habberjam. Paris: Flammarion, 1977.

Dembicki, Geoff. "A Debate Over Racism Has Spread to One of the World's Most Famous Climate Groups." *Vice*, April 28, 2020. https://www.vice.com.

Depew, David J., and John Lyne. "The Productivity of Scientific Rhetoric." *Poroi* 9, no. 1 (2013): 1–20.

Douglas, Mary, and Aaron B. Wildavsky. *Risk and Culture: An Essay on the Selection of Technical and Environmental Dangers*. Berkeley: University of California Press, 1982.

Druschke, Caroline Gottschalk. "A Tropic Future for Rhetoric Ecologies." enculturation, February 20, 2019. http://www.enculturation.net.

Dunlap, R. E., and A. M. McCright. "Organized Climate Change Denial." In *The Oxford*

Handbook of Climate Change and Society, edited by John S. Dryzek et al., 144–60. Oxford: Oxford University Press, 2011.

Dyson, Freeman J. "Can We Control the Carbon Dioxide in the Atmosphere?" *Energy* 2, no. 3 (September 1977): 287–91.

Edwards, Paul N. *A Vast Machine: Computer Models, Climate Data, and the Politics of Global Warming*. Cambridge, MA: MIT Press, 2010.

Ehrlich, Paul R. *The Population Bomb*. New York: Ballantine Books, 1968.

Elias, Amy J., and Christian Moraru, eds. *The Planetary Turn: Relationality and Geoaesthetics in the Twenty-First Century*. Evanston, IL: Northwestern University Press, 2015.

Ellerman, A. Denny, Paul L. Joskow, Richard Schmalensee, Juan-Pablo Montero, and Elizabeth M. Bailey. *Markets for Clean Air: The Acid Rain Program*. Cambridge, UK: Cambridge University Press, 2000.

Ellis, Erle C. "The Anthropocene: A New Epoch of Geological Time?" *Philosophical Transactions: Mathematical, Physical and Engineering Sciences* 369, no. 1938 (2011): 1010–35.

———. *Anthropocene: A Very Short Introduction*. Oxford: Oxford University Press, 2018.

———. "The Planet of No Return: Human Resilience on an Artificial Earth." *Breakthrough Journal* 2 (Fall 2011): 37–44.

Ellsworth, Elizabeth, and Jamie Kruse. "Evidence: Making a Geological Turn in Cultural Awareness." In *Making the Geologic Now: Responses to Material Conditions of Contemporary Life*, edited by Elizabeth Ellsworth and Jamie Kruse, 6–12. Santa Barbara, CA: Punctum Books, 2013.

Environmental Defense Fund. "Our Position on Geoengineering." Accessed June 3, 2020. https://www.edf.org/.

Epstein, Richard A. "A Climate Change Emergency?" Hoover Institution, October 7, 2019. http://www.hoover.org.

Esbjörn-Hargens, Sean. "An Ontology of Climate Change: Integral Pluralism and the Enactment of Multiple Objects." *Journal of Integral Theory and Practice* 5, no. 1 (2010): 143–74.

ETC Group. "Informational Backgrounder on the 2012 Haida Gwaii Iron Dump." March 3, 2013. https://www.etcgroup.org.

Extinction Rebellion. Accessed June 24, 2020. https://extinctionrebellion.us.

Extinction Rebellion America. "Who We Are." Accessed June 24, 2020. https://xramerica.org.

Fagan, Brian M. *The Long Summer: How Climate Changed Civilization*. New York: Basic Books, 2004.

Fahnestock, Jeanne. "Rhetoric of Science: Enriching the Discipline." *Technical Communication Quarterly* 14, no. 3 (2005): 277–86.

———. "The Rhetorical Arts of Cooperation." *JGE: The Journal of General Education* 62, no. 1 (2013): 14.

Fairbrother, Malcolm. "Geoengineering, Moral Hazard, and Trust in Climate Science:

Evidence from a Survey Experiment in Britain." *Climatic Change* 139, nos. 3–4 (December 2016): 477–89.

Ferraro, Angus J., Andrew J. Charlton-Perez, and Eleanor J. Highwood. "A Risk-Based Framework for Assessing the Effectiveness of Stratospheric Aerosol Geoengineering." *PLoS ONE* 9, no. 2 (2014): e88849.

Fleming, James Rodger. *Fixing the Sky: The Checkered History of Weather and Climate Control.* New York: Columbia University Press, 2010.

Flynn, Elizabeth A., Patricia Sotirin, and Ann Brady, eds. *Feminist Rhetorical Resilience.* Logan: Utah State University Press, 2012.

Folke, Carl, Reinette Biggs, Albert V. Norström, Belinda Reyers, and Johan Rockström. "Social-Ecological Resilience and Biosphere-Based Sustainability Science." *Ecology and Society* 21, no. 3 (September 2016): 1–16.

Folland, C. K., T. R. Karl, J. R. Christy, et al. "Observed Climate Variability and Change." In *Climate Change 2001: The Scientific Basis. Contribution of Working Group I to the Third Assessment Report of the Intergovernmental Panel on Climate Change*, edited by J. T. Houghton, Y. Ding, D. J. Griggs, et al., 100–181. Cambridge, UK: Cambridge University Press, 2001.

Frontline. Episode 8, "Plastic Wars." Produced by Rick Young, Laura Sullivan, Emma Schwartz, and Fritz Kramer. Aired March 31, 2020, on PBS.

Frost, Erin A. "Transcultural Risk Communication on Dauphin Island: An Analysis of Ironically Located Responses to the Deepwater Horizon Disaster." *Technical Communication Quarterly* 22, no. 1 (2013): 50–66.

Fuller, Steve. "The Strong Program in the Rhetoric of Science." In *Science, Reason and Rhetoric*, edited by Henry Krips, J. E. McGuire, and Trevor Melia, 95–117. Pittsburgh, PA: University of Pittsburgh Press, 1995.

Gannon, Kate Elizabeth, and Mike Hulme. "Geoengineering at the 'Edge of the World': Exploring Perceptions of Ocean Fertilisation through the Haida Salmon Restoration Corporation." *Geo: Geography and Environment* 5, no. 1 (January–June 2018): 1–25.

Gardiner, Stephen. *A Perfect Moral Storm: The Ethical Tragedy of Climate Change.* Oxford: Oxford University Press, 2013.

Gilbert, Patricia M., Rhodora Azanza, Michele Burford, et al. "Ocean Urea Fertilization for Carbon Credits Poses High Ecological Risks." *Marine Pollution Bulletin* 56, no. 6 (June 2008): 1049–56.

Goes, Marlos, Nancy Tuana, and Klaus Keller. "The Economics (or Lack Thereof) of Aerosol Geoengineering." *Climatic Change* 109, nos. 3–4 (December 2011): 719–44.

Goldenberg, Suzanne. "US Senate Refuses to Accept Humanity's Role in Global Climate Change, Again." *Guardian*, January 22, 2015. https://www.theguardian.com.

Grabill, Jeffrey T., and W. Michele Simmons. "Toward a Critical Rhetoric of Risk Communication: Producing Citizens and the Role of Technical Communicators." *Technical Communication Quarterly* 7, no. 4 (1998): 415–41.

Graeber, Charles. "Dumping Iron." *Wired*, November 1, 2000. https://www.wired.com.

Graham, S. Scott, and Carl Herndl. "Multiple Ontologies in Pain Management: Toward a Postplural Rhetoric of Science." *Technical Communication Quarterly* 22, no. 2 (2013): 103–21.

Grant, David. "Writing *Wakan*: The Lakota Pipe as Rhetorical Object." *College Composition and Communication* 69, no. 1 (2017): 61–86.

Great Barrier Reef Marine Park Authority. Accessed August 13, 2020. https://www.gbrmpa.gov.au.

Greenpeace USA. *Carbon Capture Scam (CCS): How a False Climate Solution Bolsters Big Oil.* Washington, DC: Greenpeace USA, 2015.

Gries, Laurie E. "Iconographic Tracking: A Digital Research Method for Visual Rhetoric and Circulation Studies." *Computers and Composition* 30, no. 4 (2013): 332–48.

———. "On Rhetorical Becoming." In Barnett and Boyle, *Rhetoric, through Everyday Things*, 155–70.

———. *Still Life with Rhetoric: A New Materialist Approach for Visual Rhetorics*. Logan: Utah State University Press, 2015.

Gries, Laurie E., and Collin Gifford Brooke. *Circulation, Writing, and Rhetoric*. Logan: Utah State University Press, 2018.

Gross, Alan G. "Rhetoric of Science." In *Encyclopedia of Rhetoric and Composition*, edited by Theresa Enos, 622–27. London: Routledge, 1996.

———. "Rhetoric of Science 'Is' Epistemic Rhetoric." *Quarterly Journal of Speech* 76, no. 3 (1990): 304–6.

Gruber, David R., and Randy Allen Harris. "Scientific Futures for a Rhetoric of Science: 'We Do This and They Do That?' A Junior-Senior Scholar Session, RSA 2018, Minneapolis, Minnesota, USA; 1 June 2018." *Poroi* 14, no. 2 (2019): 1–13.

Guillaume, Bertrand. "Vernadsky's Philosophical Legacy: A Perspective from the Anthropocene." *Anthropocene Review* 1, no. 2 (August 2014): 137–46.

Haas, Angela. "Wampum as Hypertext: An American Indian Intellectual Tradition of Multimedia Theory and Practice." *Studies in American Indian Literatures* 19, no. 4 (2007): 77–100.

Haida Salmon Restoration Corporation. "Ocean Micro Nutrient Replenishment." November 2, 2018. https://www.haidasalmonrestoration.com.

———. "Our Mission." Accessed March 2, 2020. http://www.haidasalmonrestoration.com/.

Hale, Erin. "Geoengineering Experiment Cancelled Due to Perceived Conflict of Interest." *Guardian*, May 16, 2012. https://www.theguardian.com.

Hamblin, Jacob Darwin. *Arming Mother Nature: The Birth of Catastrophic Environmentalism*. Oxford: Oxford University Press, 2013.

Hamilton, Clive. "The Clique That Is Trying to Frame the Global Geoengineering Debate." *Guardian*, December 5, 2011. https://www.theguardian.com.

———. *Earthmasters: The Dawn of the Age of Climate Engineering*. New Haven: Yale University Press, 2013.

———. "The Technofix Is In: A Critique of 'An Ecomodernist Manifesto.'" April 28, 2015. http://clivehamilton.com.

Hamilton, Clive, and Jacques Grinevald. "Was the Anthropocene Anticipated?" *Anthropocene Review* 2, no. 1 (April 2015): 59–72.

"Hands Off Mother Earth! Manifesto against Geoengineering." *Geoengineering Monitor*, October 4, 2018. https://www.geoengineeringmonitor.org.

Hanna, Philippe, Frank Vanclay, and Jos Arts. "The Communication and Management of Social Risks and Their Relevance to Corporate-Community Relationships." In Crichton, Candlin, and Firkins, *Communicating Risk*, 171–88.

Hansen, James, Masahiro Sato, Pushker Kharecha, et al. "Target Atmospheric CO_2: Where Should Humanity Aim?" *Open Atmospheric Science Journal* 2, no. 1 (2008): 217–31.

Haraway, Donna. "Situated Knowledges: The Science Question in Feminism and the Privilege of Partial Perspective." *Feminist Studies* 14, no. 3 (1988): 575–99.

Harris, Randy Allen. "Knowing, Rhetoric, Science." In *Visions and Revisions: Continuity and Change in Rhetoric and Composition*, edited by James D. Williams, 162–218. Carbondale: Southern Illinois University Press, 2002.

Hausfather, Zeke. "Mission 2020: A New Global Strategy to 'Rapidly' Reduce Carbon Emissions." CarbonBrief, June 28, 2017. https://www.carbonbrief.org.

Hawhee, Debra. *Bodily Arts: Rhetoric and Athletics in Ancient Greece*. Austin: University of Texas Press, 2004.

Hawk, Byron. *A Counter-History of Composition: Toward Methodologies of Complexity*. Pittsburgh, PA: University of Pittsburgh Press, 2007.

Heede, Richard. *Carbon Majors: Updating Activity Data, Adding Entities, & Calculating Emissions: A Training Manual*. Snowmass, CO: Climate Accountability Institute, 2019.

Henriksen, Christina, Johanna Sandahl, Mikael Sundström, and Isadora Wronski. Letter to the SCoPEx Advisory Committee, February 24, 2021.

Henwood, Karen L., and Nick Pidgeon. "Interpretive Environmental Risk Research: Affect, Discourses and Change." In Crichton, Candlin, and Firkin, *Communicating Risk*, 155–70.

Herndl, Carl G., and Lauren Leigh Cutlip. "'How Can We Act?' A Praxiographical Program for the Rhetoric of Technology, Science, and Medicine." *Poroi* 9, no. 1 (2013): 1–13.

Holland, Eugene W. *Deleuze and Guattari's* A Thousand Plateaus. London: Bloomsbury, 2013.

Holling, C. S. "Engineering Resilience versus Ecological Resilience." In *Engineering within Ecological Constraints*, edited by Peter C. Schulze, 31–44. Washington, DC: National Academies Press, 1996.

———. "Resilience and Stability of Ecological Systems." *Annual Review of Ecology and Systematics* 4 (1973): 1–23.

Hoofnagel, Mark. "Climate Change Deniers: Failsafe Tips on How to Spot Them." *Guardian*, March 11, 2019. https://www.theguardian.com.

Horton, Zach. "Going Rogue or Becoming Salmon? Geoengineering Narratives in Haida Gwaii." *Cultural Critique* 97 (2017): 128–66.

Hourdequin, Marion. "Climate Change, Climate Engineering, and the 'Global Poor':

What Does Justice Require?" *Ethics, Policy and Environment* 21, no. 3 (2018): 270–88.

———. "Geoengineering, Solidarity, and Moral Risk." In *Engineering the Climate: The Ethics of Solar Radiation Management*, edited by Christopher J. Preston, 15–32. Lanham, MD: Lexington Books, 2012.

House of Commons Science and Technology Committee. "The Regulation of Geoengineering." The Stationery Office Limited House of 23, Fifth Report of Session, 2009–2010. March 10, 2010. https://publications.parliament.uk.

House Select Committee on the Climate Crisis. *Solving the Climate Crisis: The Congressional Action Plan for Clean Energy and a Healthy, Resilient, and Just America*. June 2020. https://climatecrisis.house.gov.

Hulme, Mike. *Why We Disagree about Climate Change*. Cambridge, UK: Cambridge University Press, 2009.

Huttunen, Suvi, and Mikael Hildén. "Framing the Controversial: Geoengineering in Academic Literature." *Science Communication* 36, no. 1 (February 2014): 3–29.

Iklé, Fred, and Lowell Wood. "Sunscreen for Planet Earth." *Hoover Digest* 1, 2008. https://www.hoover.org.

Indigenous Peoples' Global Summit on Climate Change. "The Anchorage Declaration." Anchorage, April 24, 2009. https://unfccc.int.

Interagency Working Group on Social Cost of Carbon. "Technical Support Document: Social Cost of Carbon for Regulatory Impact Analysis under Executive Order 12866." United States Environmental Protection Agency, 2010. www.epa.gov.

Intergovernmental Panel on Climate Change. *Climate Change 2014: Mitigation of Climate Change, The Working Group III contribution to the Fifth Assessment Report (AR5) of the Intergovernmental Panel on Climate Change*. New York: Cambridge University Press, 2014. https://www.ipcc.ch.

———. *IPCC Expert Meeting on Geoengineering, Meeting Report*. 2012. https://archive.ipcc.ch.

———. *Special Report: Global Warming of 1.5°C*. https://www.ipcc.ch.

International Energy Agency. "Global Energy Demand to Plunge This Year as a Result of the Biggest Shock since the Second World War." April 30, 2020. https://www.iea.org.

International Maritime Organization. Convention on the Prevention of Marine Pollution by Dumping Wastes and Other Matter. London, December 29, 1972. https://www.imo.org.

International Organization for Standardization. "ISO Standard no. 14082, Working Draft 2: GHG Management—Guidance for the Quantification and Reporting of Radiative Forcing Based Climate Footprint and Mitigation Efforts." https://www.desmog.com.

"Iron Fertilization." Living on Earth. Accessed June 11, 2020. https://loe.org.

Izrael, Yuri, Alexey Ryaboshapko, and N. N. Petrov. "Comparative Analysis of Geo-Engineering Approaches to Climate Stabilization." *Russian Meteorology and Hydrology* 34, no. 6 (2009): 335–47.

Izrael, Yuri, V. M. Zakharov, N. N. Petrov, et al. "Field Experiment on Studying Solar

Radiation Passing through Aerosol Layers." *Russian Meteorology and Hydrology* 34, no. 5 (2009): 265–73.

Jacques, Peter J., Riley E. Dunlap, and Mark Freeman. "The Organisation of Denial: Conservative Think Tanks and Environmental Scepticism." *Environmental Politics* 17, no. 3 (2008): 349–85.

Jasanoff, Sheila, and Sang-Hyun Kim. "Containing the Atom: Sociotechnical Imaginaries and Nuclear Power in the United States and South Korea." *Minerva* 47, no. 2 (June 2009): 119–46.

Kahan, Dan M., Hank Jenkins-Smith, Tor Tarantola, Carol L. Silva, and Donald Braman. "Geoengineering and Climate Change Polarization: Testing a Two-Channel Model of Science Communication." *Annals of the American Academy of Political and Social Science* 658 (2015): 192–222.

Kahn, Brian. "Republicans Decided to Talk about a Dramatic Climate Solution without Addressing the Actual Problem." *Gizmodo*, August 11, 2017. https://gizmodo.com.

Kahneman, Daniel. *Thinking, Fast and Slow*. New York: Farrar, Straus and Giroux, 2011.

Katz, Steven B., and Carolyn R. Miller. "The Low-Level Radioactive Waste Siting Controversy in North Carolina: Toward a Rhetorical Model of Risk Communication." In *Green Culture: Environmental Rhetoric in Contemporary America*, edited by C. G. Herndl and S. C. Brown, 111–40. Madison: University of Wisconsin Press, 1996.

Keith, David. "A Critical Look at Geoengineering against Climate Change." *TEDSalon*, 2007. https://www.ted.com.

———. "Geoengineering the Climate: History and Prospect." *Annual Review of Energy and the Environment* 25 (2000): 245–84.

Keith, David, and Hadi Dowlatabadi. "A Serious Look at Geoengineering." *Eos, Transactions of the American Geophysical Union* 73, no. 27 (1992): 289–93.

Keith, David, and Joshua Horton. "Solar Geoengineering and Obligations to the Global Poor." In *Climate Justice and Geoengineering: Ethics and Policy in the Atmospheric Anthropocene*, edited by Christopher J. Preston, 79–92. London: Row man and Littlefield, 2016.

Keith, David W., Edward Parson, and M. Granger Morgan. "Research on Global Sun Block Needed Now." *Nature* 463, no. 7280 (2010): 426–27.

Kennedy, George A. *A New History of Classical Rhetoric*. Princeton: Princeton University Press, 1994.

Kimmerer, Robin Wall. *Braiding Sweetgrass: Indigenous Wisdom, Scientific Knowledge and the Teachings of Plants*. Minneapolis: Milkweed Editions, 2013.

King, Michael D., Ian M. Howat, Salvatore G. Candela, et al. "Dynamic Ice Loss from the Greenland Ice Sheet Driven by Sustained Glacier Retreat." *Communications Earth Environment* 1, no. 1 (2020).

Kintisch, Eli. *Hack the Planet: Science's Best Hope—or Worst Nightmare—for Averting Climate Catastrophe*. Hoboken, NJ: Wiley, 2010.

Kirk, Andrew G. *Counterculture Green: The Whole Earth Catalog and American Environmentalism*. Lawrence: University Press of Kansas, 2007.

Klein, Naomi. "Geoengineering: Testing the Waters." *New York Times*, October 28, 2012. https://www.nytimes.com.

———. *This Changes Everything: Capitalism vs. the Climate*. New York: Simon and Schuster, 2014.

Kravitz, Ben, Alan Robock, Piers M. Forster, James M. Haywood, Mark G. Lawrence, and Hauke Schmidt. "An Overview of the Geoengineering Model Intercomparison Project (GeoMIP)." *Journal of Geophysical Research Atmospheres* 118, no. 23 (2013): 13103–7.

Lakoff, George. "Why It Matters How We Frame the Environment." *Environmental Communication* 4, no. 1 (2010): 70–81.

Lamb, William F., Giulio Mattioli, Sebastian Levi, et al. "Discourses of Climate Delay." *Global Sustainability* 3, e17 (2020): 1–5.

Latour, Bruno. *Pandora's Hope: Essays on the Reality of Science Studies*. Cambridge, MA: Harvard University Press, 1999.

———. *We Have Never Been Modern*. Translated by Catherine Porter. Cambridge, MA: Harvard University Press, 1993.

Law, John. *After Method: Mess in Social Science Research*. London: Routledge, 2004.

———. "Enacting Naturecultures: A View from STS." *Heterogeneities*, February 24, 2004. www.heterogeneities.net.

Leiserowitz, Anthony, Edward Maibach, Seth Rosenthal, et al. "Climate Change in the American Mind: December 2018." Yale Program on Climate Change Communication and George Mason University Center for Climate Change Communication, January 22, 2019. https://climatecommunication.yale.edu.

Leiss, William, and Douglas Powell. *Mad Cows and Mother's Milk: The Perils of Poor Risk Communication*. Montreal: McGill-Queen's University Press, 1997.

Lepori, Matthew. "There Is No Anthropocene: Climate Change, Species-Talk, and Political Economy." *Telos* 172 (Fall 2015): 103–24.

Levin, Kelly. "New Global CO_2 Emissions Numbers Are In. They're Not Good." *World Resources Institute*, December 5, 2018. https://www.wri.org.

Levitan, Dave. "For Geoengineers, a Scientific Existential Crisis." *UNDARK*, January 1, 2019. https://undark.org.

Levitt, Steven D., and Stephen J. Dubner. *SuperFreakonomics: Global Cooling, Patriotic Prostitutes, and Why Suicide Bombers Should Buy Life Insurance*. New York: HarperCollins, 2009.

Li, Jun. "Governing Urban Infrastructure in Developing Cities: The Role of Carbon Finance." In *Global Change, Energy Issues and Regulation Policies*, edited by Jean Bernard Saulnier and Marcelo D. Varella, 259–77. Heidelberg: Springer, 2013.

Lin, Albert C. "Does Geoengineering Present a Moral Hazard?" *Ecology Law Quarterly* 40, no. 3 (2013): 673–712.

Lockley, Andrew, and D'Maris Coffman. "Distinguishing Morale Hazard from Moral Hazard in Geoengineering." *Environmental Law Review* 18, no. 3 (2016): 194–204.

Lopez, Gabriel, Zack Abbott, Greg Rau, Leonid Kozhukh, and Sam Altman. "Carbon Removal Technologies." *Y Combinator*. Accessed April 4, 2021. https://www.ycombinator.com.

Lovelock, James E., and Chris G. Rapley. "Ocean Pipes Could Help the Earth to Cure Itself." *Nature* 449, no. 403 (2007): 403.

Lowman, Meg. "Geo-Engineering a Rescue?" *Herald-Tribune*, August 23, 2010. https://www.heraldtribune.com.

Lucht, W., and R. K. Pachauri. "The Mental Component of the Earth System." In *Earth System Analysis for Sustainability*, edited by Hans Joachim Schellnhuber et al., 341–66. Cambridge, MA: MIT Press, 2004.

Lukacs, Martin. "World's Biggest Geoengineering Experiment 'Violates' UN Rules." *Guardian*, October 15, 2012. https://www.theguardian.com.

Luokkanen, Matti, Suvi Huttunen, and Mikael Hildén. "Geoengineering, News Media and Metaphors: Framing the Controversial." *Public Understanding of Science* 23, no. 8 (November 2014): 966–81.

Lupton, Deborah. *Risk*. London: Routledge, 1999.

Lynas, Mark. *The God Species: Saving the Planet in the Age of Humans*. Washington, DC: National Geographic, 2011.

MacCracken, Michael, Paul Berg, Paul Crutzen, et al. "Conference Statement." Asilomar International Conference on Climate Intervention Technologies, Pacific Grove, CA, 22–26 March 2010. http://climate.org.

Malm, Andreas. *Fossil Capital: The Rise of Steam Power and the Roots of Global Warming*. London: Verso, 2016.

Malm, Andreas, and Alf Hornborg. "The Geology of Mankind? A Critique of the Anthropocene Narrative." *Anthropocene Review* 1, no. 1 (April 2014): 62–69.

Mann, Geoff, and Joel Wainwright. *Climate Leviathan: A Political Theory of Our Planetary Future*. London: Verso, 2018.

Mann, Charles C., "The Book That Incited a Worldwide Fear of Overpopulation: 'The Population Bomb' Made Dire Predictions—and Triggered a Wave of Repression Around the World," *Smithsonian Magazine*, January 2018

Mann, Michael E. *The Hockey Stick and the Climate Wars: Dispatches from the Front Lines*. New York: Columbia University Press, 2012.

Marback, Richard. "Unclenching the Fist: Embodying Rhetoric and Giving Objects Their Due." *Rhetoric Society Quarterly* 38, no. 1 (2008): 46–65.

Marchetti, Cesare. "On Geoengineering and the CO_2 Problem." *Climatic Change* 1, no. 1 (1977): 59–68.

Marshall, George. *Don't Even Think about It: Why Our Brains Are Wired to Ignore Climate Change*. New York: Bloomsbury, 2014.

Martin, John H., and Steve E. Fitzwater. "Iron Deficiency Limits Phytoplankton Growth in the North-East Pacific Subarctic." *Nature* 331, no. 6154 (1988): 341–43.

Martin, John H., Steve E. Fitzwater, and R. Michael Gordon. "Iron Deficiency Limits Phytoplankton Growth in Antarctic Waters." *Global Biogeochemical Cycles* 4, no. 1 (March 1990): 5–12.

Matthews, Damon H., and Ken Caldeira. "Transient Climate Carbon Simulations of Planetary Geoengineering." *PNAS* 104, no. 24 (June 2007): 9949–54.

McGrath, Matt. "Key 'Step Forward' in Cutting Cost of Removing CO_2 from Air." *BBC News*, June 7, 2018. https://www.bbc.com.

McGreavy, Bridie, Justine Wells, George F. McHendry Jr., and Samantha Senda-Cook, eds. *Tracing Rhetoric and Material Life: Ecological Approaches*. London: Palgrave Macmillan, 2018.

McKell, Sue, and Paul De Barro. "Between Two Absolutes Lies Risk: Risk Communication in Biosecurity Discourse." In Crichton, Candlin, and Firkin, *Communicating Risk*, 229–44.

McKibben, Bill. *The End of Nature*. New York: Random House, 2006.

McKinnon, Catriona. "Sleepwalking into Lock-In? Avoiding Wrongs to Future People in the Governance of Solar Radiation Management Research." *Environmental Politics* 28, no. 3 (2019): 441–59.

Merchant, Carolyn. *The Death of Nature: Women, Ecology, and the Scientific Revolution*. New York: HarperCollins, 1980.

Merk, Christine, Gert Pönitzsch, and Katrin Rehdanz. "Knowledge about Aerosol Injection Does Not Reduce Individual Mitigation Efforts." *Environmental Research Letters* 11, no. 5 (2016): 1–6.

Micciche, Laura R. "Writing Material." *College English* 76, no. 6 (2014): 488–505.

Mignolo, Walter. *The Darker Side of Western Modernity: Global Futures, Decolonial Options*. Durham, NC: Duke University Press, 2011.

Milman, Oliver. "James Hansen, Father of Climate Change Awareness, Calls Paris Talks 'a Fraud.'" *Guardian*, December 12, 2015. https://www.theguardian.com.

Mol, Annemarie. *The Body Multiple: Ontology in Medical Practice*. Durham, NC: Duke University Press, 2003.

———. "Ontological Politics: A Word and Some Questions." Supplement, *Sociological Review* 47, no. 1 (1999): 74–89.

Moore, Jason W. "Introduction: Anthropocene or Capitalocene? Nature, History, and the Crisis of Capitalism." In Moore, *Anthropocene or Capitalocene?*, 1–13.

———. "The Rise of Cheap Nature." In Moore, *Anthropocene or Capitalocene?*, 78–115.

Moore, Jason W., ed. *Anthropocene or Capitalocene? Nature, History, and the Crisis of Capitalism*. Oakland, CA: PM Press, 2016.

Moraru, Christian. *Reading for the Planet: Toward a Geomethodology*. Ann Arbor: University of Michigan Press, 2015.

Morozov, Evgeny. *To Save Everything, Click Here: The Folly of Technological Solutionism*. New York: Public Affairs, 2013.

Morton, Timothy. *The Ecological Thought*. Cambridge, MA: Harvard, University Press, 2010.

———. *The Planet Remade: How Geoengineering Could Change the World*. Princeton, NJ: Princeton University Press, 2010.

Murphy, Robert P. "The Benefits of Procrastination: The Economics of Geo-Engineering." Library of Economics and Liberty, December 7, 2009. https://www.econlib.org.

National Research Council. *Advancing the Science of Climate Change*. Washington, DC: National Academies Press, 2010.

———. *Basic Research Opportunities in Earth Science*. Washington, DC: National Academies Press, 2001.

Neyrat, Frederic. *The Unconstructable Earth: An Ecology of Separation*. New York: Fordham University Press, 2018.

Nixon, Rob. *Slow Violence and the Environmentalism of the Poor*. Cambridge, MA: Harvard University Press, 2013.

Nordhaus, Ted, and Michael Shellenberger. "The Death of Environmentalism: Global Warming Politics in a Post-Environmental World." Breakthrough Institute, September 24, 2004. https://thebreakthrough.org.

Nyks, Kelly, and Jared P. Scott, dir. *Disruption*. 2014. PF Pictures.

Ocean Nourishment Corporation. "Agenda: Waste Atmospheric Carbon Dioxide." Accessed April 14, 2020. www.oceannourishment.com.

Oreskes, Naomi, and Erik M. Conway. *Merchants of Doubt: How a Handful of Scientists Obscured the Truth on Issues from Tobacco Smoke to Global Warming*. New York: Bloomsbury Press, 2010.

Oreskes, Naomi, Kristin Shrader-Frenchette, and Kenneth Belitz. "Verification, Validation, and Confirmation of Numerical Models in the Earth Sciences." *Science* 263, no. 5147 (1994): 641–46.

Paine, R. T. "A Note on Trophic Complexity and Community Stability." *American Naturalist* 103, no. 929 (1969): 91–93.

Parikka, Jussi. *A Geology of Media*. Minneapolis: University of Minnesota Press, 2015.

Parliament of Canada. Canadian Environmental Protection Act. 1999, SC 1999, c 33, https://canlii.ca.

Parmenides of Elea. *Fragments*. Translated by David Gallop. Toronto: University of Toronto Press, 2013.

Parson, Edward, A. "Reflections on Air Capture: The Political Economy of Active Intervention in the Global Environment." *Climatic Change* 74, no. 1 (2006): 5–15.

Pellizzoni, Luigi. *Ontological Politics in a Disposable World: The New Mastery of Nature*. London: Routledge, 2015.

Pender, Kelly. *Techne: From Neoclassicism to Postmodernism: Understanding Writing as a Useful, Teachable Art*. Clemson, SC: Parlor Press, 2011.

Perelman, Chaïm, and L. Olbrechts-Tyteca. *The New Rhetoric: A Treatise on Argumentation*. Translated by John Wilkinson and Purcell Weaver. South Bend, IN: University of Notre Dame Press, 1973.

Pflugfelder, Ehren Helmut. *Communicating Mobility and Technology: A Material Rhetoric for Persuasive Transportation*. London: Routledge, 2016.

———. "Failure Matters: Conflicting Practices in a High-Tech Case." *Journal of Technical Writing and Communication* 48, no. 1 (January 2018): 31–52.

———. "Rhetoric's New Materialism: From Micro-Rhetoric to Microbrew." *Rhetoric Society Quarterly* 45, no. 5 (2015): 441–61.

Potts, Liza. *Social Media in Disaster Response: How Experience Architects Can Build for Participation*. London: Routledge, 2014.

Preiser, Rika, Laura M. Pereira, and Reinette (Oonise) Biggs. "Navigating Alternative Framings of Human-Environment Interactions: Variations on the Theme of 'Finding Nemo.'" *Anthropocene* 20 (December 2017): 83–87.

Prelli, Lawrence J. "Stasis and the Problem of Incommensurate Communication: The

Case of Spousal Violence Research." In *Rhetoric and Incommensurability*, edited by Randy Allen Harris, 294–333. West Lafayette, IN: Parlor Press, 2005.

President's Science Advisory Committee, Environmental Pollution Panel. *Restoring the Quality of Our Environment*. Washington, DC: White House, 1965. http://www.climatefiles.com.

Preston, Christopher J. *The Synthetic Age: Outdesigning Evolution, Resurrecting Species, and Reengineering Our World*. Cambridge, MA: MIT Press, 2018.

Propen, Amy D. *Visualizing Posthuman Conservation in the Age of the Anthropocene*. Columbus: Ohio State University Press, 2018.

Rai, Candice, and Caroline Gottschalk Druschke. *Field Rhetoric: Ethnography, Ecology, and Engagement in the Places of Persuasion*. Tuscaloosa: University of Alabama Press, 2018.

Rakestraw, Norris W., and Charles D. Keeling. "Proposal to the National Science Foundation for A Study of the Abundance of CO_2." Scripps Institution of Oceanography, folder "Government Correspondence—NSF - 1958–1969," Keeling office files, SIO, June 1958.

Rasch, Philip, Simone Tilmes, Richard P. Turco, et al. "An Overview of Geoengineering of Climate using Stratospheric Sulfate Aerosols." *Philosophical Transactions of the Royal Society A* 366, no. 1882 (2008): 4007–37.

Reef Restoration and Adaptation Program. "Winning Hearts, Finding Cash, and Tough Decisions to Save the Reef." July 18, 2018. https://gbrrestoration.org.

Revkin, Andrew C. "Managing Planet Earth; Forget Nature. Even Eden Is Engineered." *New York Times*, August 20, 2002. https://www.nytimes.com.

Reynolds, Jesse. "A Critical Examination of the Climate Engineering Moral Hazard and Risk Compensation Concern." *Anthropocene Review* 2, no. 2 (August 2015): 174–91.

———. "A Solar Geoengineering Milestone Goes Largely Unnoticed." LegalPlanet, May 6, 2020. https://legal-planet.org.

Rice, Jeff. *Networked Detroit: Rhetoric and Space in the Age of the Network*. Carbondale: Southern Illinois University Press, 2012.

Rickert, Thomas. "Afterword: A Crack in the Cosmic Egg, Tuning into Things." In Barnett and Boyle, *Rhetoric, through Everyday Things*, 226–32.

———. *Ambient Rhetoric: The Attunements of Rhetorical Being*. Pittsburgh, PA: University of Pittsburgh Press, 2013.

———. "The Whole of the Moon: Latour, Context, and the Problem of Holism." In *Thinking with Bruno Latour in Rhetoric and Composition*, edited by Paul Lynch and Nathaniel Rivers, 135–50. Carbondale: Southern Illinois University Press, 2015.

Rìos, Gabriela Raquel. "Cultivating Land-Based Literacies and Rhetorics." *Literacy in Composition Studies* 3, no. 1 (2015): 60–70.

Rivers, Nathaniel A. "Deep Ambivalence and Wild Objects: Toward a Strange Environmental Rhetoric." *Rhetoric Society Quarterly* 45, no. 5 (2015): 420–40.

Robock, Alan. "20 Reasons Why Geoengineering May Be a Bad Idea." *Bulletin of the Atomic Scientists* 64, no. 2 (2008): 14–18.

Robock, Alan, Allison Marquardt, Ben Kravitz, and Georgiy Stenchikov. "Benefits,

Risks, and Costs of Stratospheric Geoengineering." *Geophysical Research Letters* 36, no. 19 (October 2009): 1–9.

———. "Regional Climate Responses to Geoengineering with Tropical and Arctic SO_2 Injections." *Journal of Geophysical Research* 113, no. D16 (2008): 1–15.

Robock, Alan, Douglas G. MacMartin, Riley Duren, and Matthew W. Christensen. "Studying Geoengineering with Natural and Anthropogenic Analogs." In "Geoengineering Research and Its Limitations." Special issue, *Climatic Change* 121, no. 3 (December 2013): 445–58.

Rohde, Robert (@RARohde). "Carbon Dioxide Emission Scenarios for 1.5°C of Warming." *Twitter*, October 13, 2018. https://twitter.com.

Romm, Joe. "Geoengineering and the New Climate Denialism." *Think Progress*, April 29, 2009. https://thinkprogress.org.

Roochnik, David. *Of Art and Wisdom: Plato's Understanding of Techne*. State College: Penn State University Press, 1996.

Rosiek, Jerry Lee, Jimmy Snyder, and Scott L. Pratt. "The New Materialisms and Indigenous Theories of Non-Human Agency: Making the Case for Respectful Anti-Colonial Engagement." *Qualitative Inquiry* 26, nos. 3–4 (March–April 2020): 331–46.

Roundtree, Aimee Kendall. *Computer Simulation, Rhetoric, and the Scientific Imagination: How Virtual Evidence Shapes Science in the Making and in the News*. Lanham, MD: Lexington Books, 2017.

The Royal Society. *Geoengineering the Climate: Science, Governance and Uncertainty*. London: The Royal Society, 2009.

Salter, Stephen, Graham Sortino, and John Latham. "Sea-Going Hardware for the Cloud Albedo Method for Reversing Global Warming." *Philosophical Transactions of the Royal Society A* 336, no. 1882 (November 2008): 3989–4006.

Samson, Paul R., and David Pitt, ed. *The Biosphere and Noosphere Reader: Global Environment, Society, and Change*. London: Routledge, 2012.

Sandmo, Agnar. "The Early History of Environmental Economics." *Review of Environmental Economics and Policy* 9, no. 1 (Winter 2015): 43–63.

Sanz-Pérez, Eloy S., Christopher R. Murdock, Stephanie A. Didas, and Christopher W. Jones. "Direct Capture of CO_2 from Ambient Air." *Chemical Reviews* 116, no. 19 (2016): 11840–76.

Satterthwaite, David. "The Implications of Population Growth and Urbanization for Climate Change." *Environment and Urbanization* 21, no. 2 (October 2009): 545–67.

Sauer, Beverly A. *The Rhetoric of Risk: Technical Documentation in Hazardous Environments*. London: Routledge, 2002.

Saunders, Harry. "Does Capitalism Require Endless Growth?" Breakthrough Institute, June 21, 2016. https://thebreakthrough.org.

Schadewaldt, Wolfgang. "The Greek Concepts of 'Nature' and 'Technique.'" In *Philosophy of Technology: The Technological Condition; An Anthology*, edited by Robert C. Scharff and Val Dusek, 2nd ed., 25–32. London: Wiley Blackwell, 2014.

Scheer, Dirk, and Ortwin Renn. "Public Perception of Geoengineering and Its

Consequences for Public Debate." *Climatic Change* 125, nos. 3–4 (August 2014): 305–18.

Schelling, Wilhelm Joseph von. *First Outline of a System of the Philosophy of Nature*. Translated by Keith R. Peterson. Albany: SUNY Press, 2004.

———. *Ideas for a Philosophy of Nature*. Translated by Errol E. Harris and Peter Heath. Cambridge, UK: Cambridge University Press, 2001.

SCoPEx Advisory Committee. "The SCoPEx Advisory Committee Is Updating and Diversifying Its Membership." April 8, 2021. https://scopexac.com.

Scott, Dane. "Philosophy of Technology and Geoengineering (Working Paper)." *Geoengineering Our Climate?* (Blog), April 23, 2013. https://geoengineeringourclimate .wordpress.com.

Scott, Heidi. "Colonialism, Landscape and the Subterranean." *Geography Compass* 2, no. 6 (November 2008): 1853–69.

Scott, J. Blake. "Kairos as Indeterminate Risk Management: The Pharmaceutical Industry's Response to Bioterrorism." *Quarterly Journal of Speech* 92, no. 2 (2006): 115–43.

Scripps Institute of Oceanography. "The Keeling Curve." Accessed July 1, 2019. https:// keelingcurve.ucsd.edu.

Seitz, Russell. "Bright Water: Hydrosols, Water Conservation and Climate Change." *Climatic Change* 105, nos. 3–4 (April 2011): 365–81.

Self, Stephen, Jin-Xia Zhao, Rick E. Holasek, Ronnie C. Torres, and Alan J. King. "The Atmospheric Impact of the 1991 Mount Pinatubo Eruption." In *Fire and Mud: Eruptions and Lahars of Mount Pinatubo, Philippines*, edited by C. G. Newhall and R. S. Punongbayan, 1089–116. Seattle: University of Washington Press, 1996.

Sheppard, B., M. Janoske, and B. Liu. *Understanding Risk Communication Theory: A Guide for Emergency Managers and Communicators*. Report to Human Factors/ Behavioral Sciences Division, Science and Technology Directorate, US Department of Homeland Security. College Park, MD, 2012.

Sikina, Jinnah, Simon Nicholson, and Jane Flegal. "Toward Legitimate Governance of Solar Geoengineering Research: A Role for Sub-State Actors." *Ethics, Policy & Environment* 21, no. 3 (2018): 362–81.

Sikka, Tina. "A Critical Discourse Analysis of Geoengineering Advocacy." *Critical Discourse Studies* 9, no. 2 (2012): 163–75.

———. "Geoengineering in a World Risk Society." *International Journal of Climate Change: Impacts and Responses* 3, no. 1 (2012): 143–54.

Simmons, W. Michele. *Participation and Power: Civic Discourse in Environmental Policy Decisions*. Albany, NY: SUNY Press, 2007.

Slovic, Paul, and Elke U. Weber. "Perception of Risk Posed by Extreme Events." Paper discussed at the conference "Risk Management Strategies in an Uncertain World." Palisades, New York, April 12–13, 2002.

Socolow, Robert, Michael Desmond, Robert Aines, et al. *Direct Air Capture of CO_2 with Chemicals*. Report for the American Physical Society Panel on Public Affairs, June 1, 2011. https://www.aps.org.

Southern Cross University. "Scientists Trial World-First 'Cloud Brightening' Technique to Protect Corals." April 17, 2020. https://www.scu.edu.au.
Specter, Michael. "The Climate Fixers." *New Yorker*, May 7, 2012. https://www.newyorker.com.
Spivak, Gayatri Chakravorty. *Death of a Discipline*. New York: Columbia University Press, 2003.
Staggers, Julie, and Michelle Simmons. "Secret/Agent." In *Posthuman Praxis in Technical Communication*, edited by Kristen R. Moore and Daniel P. Richards, 46–66. London: Routledge, 2018.
Steffen, Will, Jacques Grinevald, Paul Crutzen, and John McNeill. "The Anthropocene: Conceptual and Historical Perspectives." *Philosophical Transactions of the Royal Society* 369, no. 1938 (2011): 842–67.
Stenmark, Mikael. "What Is Scientism?" *Religious Studies* 33, no. 1 (1997): 15–32.
Stern, Nicholas. *The Economics of Climate Change: The Stern Review*. Cambridge, UK: Cambridge University Press, 2007.
Stormer, Nathan, and Bridie McGreavy. "Thinking Ecologically about Rhetoric's Ontology: Capacity, Vulnerability, and Resilience." *Philosophy and Rhetoric* 50, no. 1 (2017): 1–25.
Strathern, Marilyn. "No Nature, No Culture: The Hagen Case." In *Nature, Culture and Gender*, edited by Carol MacCormack and Marilyn Strathern, 174–222. Cambridge, UK: Cambridge University Press, 1980.
Strong, Aaron L., John J. Cullen, and Sallie W. Chisholm. "Ocean Fertilization: Science, Policy, and Commerce." *Oceanography* 22, no. 3 (2009): 236–61.
Suess, Eduard. *The Face of the Earth / Das Antlitz der Erde*. Oxford: Clarendon Press, 1904.
Swyngedouw, Erik. "Impossible Sustainability and the Post-Political Condition." In *Making Strategies in Spatial Planning: Knowledge and Values*, edited by Maria Cerreta, Grazia Concilio, and Valeria Monno, 185–205. London: Springer, 2010.
———. "The Violence of Sustainable Urbanity." *Harvard Design Magazine* 37 (Winter 2014): 24–29.
Sydell, Laura and Alex Blumberg. "When Patents Attack!" This American Life, July 22, 2011. WBEZ Chicago / PRX The Public Radio Exchange. https://www.thisamericanlife.org/.
Szeman, Imre. *After Oil*. Morgantown: West Virginia University Press, 2016.
———. *Energy Culture: Art and Theory on Oil and Beyond*. Morgantown: West Virginia University Press, 2019.
———. "Entrepreneurship as the New Common Sense." *South Atlantic Quarterly* 114, no. 3 (July 2015): 471–90.
———. *On Petrocultures: Globalization, Culture, Energy*. Morgantown: West Virginia University Press, 2019.
Szeman, Imre, and Dominic Boyer. "Introduction: On the Energy Humanities." In *Energy Humanities: An Anthology*, edited by Imre Szeman and Dominic Boyer, 15–46. Baltimore: Johns Hopkins University Press, 2017.
Szersynski, Bronislaw, Matthew Kearnes, Phil Macnaghten, Richard Owen, and Jack

Stilgoe. "Why Solar Radiation Management Geoengineering and Democracy Won't Mix." *Environment and Planning* 45, no. 12 (December 2013): 2809–16.

TallBear, Kim. "Beyond the Life/Not Life Binary: A Feminist-Indigenous Reading of Cryopreservation, Interspecies Thinking and the New Materialisms." In *Cryopolitics: Frozen Life in a Melting World*, edited by Joanna Radin and Emma Kowal, 179–202. Cambridge, MA: MIT Press, 2017.

Tans, Pieter, and Kirk Thoning. "How We Measure Background CO_2 Levels on Mauna Loa." NOAA Earth System Research Laboratory. Accessed March 13, 2019. https://gml.noaa.gov.

Teller, E., L. Wood, and R. Hyde. "Global Warming and Ice Ages: I. Prospects for Physics-Based Modulation of Global Change." International Atomic Energy Agency, 1997. https://mronline.org.

Thernstrom, Samuel. "What Role for Geoengineering?" American Enterprise Institute, March 2, 2020. https://www.aei.org.

Tollefson, Jeff. "First Sun-Dimming Experiment Will Test a Way to Cool Earth." *Nature*, November 27, 2018. https://www.nature.com.

Tranchard, Sandrine. "The New ISO 31000 Keeps Risk Management Simple." International Organization for Standardization, February 15, 2018. https://www.iso.org.

Tschakert, Petra. "Whose Hands Are Allowed at the Thermostat?" Paper presented at the conference titled "The Ethics of Geoengineering: Investigating the Moral Challenges of Solar Radiation Management, 2010," University of Montana, Missoula, MO, October 8–20, 2010. http://www.umt.edu.

Turner, Fred. *From Counterculture to Cyberculture: Stewart Brand, the Whole Earth Network, and the Rise of Digital Utopianism*. Chicago: University of Chicago Press, 2010.

United Nations. Cartagena Protocol on Biosafety to the Convention on Biological Diversity. Cartagena, May 16, 2000. No. 30619. https://bch.cbd.int.

———. Convention on Biological Diversity. Rio de Janeiro, June 5, 1992. No. 30619. https://www.cbd.int.

———. Convention on Biological Diversity. "Climate-Related Geoengineering and Biodiversity: Technical and Regulatory Matters on Geoengineering in Relation to the CBD." March 23, 2017. https://www.cbd.int.

———. Convention on the Prohibition of Military or Any Other Hostile Use of Environmental Modification Technique. New York, December 10, 1976. No. 17119. https://treaties.un.org.

———. Framework Convention on Climate Change. New York, May 9, 1992. No. 30822. https://treaties.un.org.

United Nations Economic Condition for Europe. Convention on Long-Range Transboundary Air Pollution. Geneva, November 13, 1979. No. 21623. https://treaties.un.org.

United Nations Environment Programme. *Decoupling 2: Technologies, Opportunities and Policy Options. A Report of the Working Group on Decoupling to the International Resource Panel*, 2014. https://www.unep.org.

———. The Montreal Protocol on Substances That Deplete the Ozone Layer. Montreal, September 15, 1987. https://www.unep.org.

———. The Vienna Convention for the Protection of the Ozone Layer. Vienna, March 22, 1985. https://ozone.unep.org.

United Nations Office for Outer Space Affairs. Treaty on Principles Governing the Activities of States in the Exploration and Use of Outer Space, including the Moon and Other Celestial Bodies. New York, January 27, 1967. Resolution 2222 (XXI). https://www.unoosa.org.

United States Government Accountability Office. *Climate Engineering: Technical Status, Future Directions, and Potential Responses*. Report to Congressional Requester, GAO-11-71, July 2011. https://www.gao.gov.

Urry, John. *Mobilities*. London: Polity, 2007.

Van der Tuin, Iris. "A Different Starting Point, a Different Metaphysics: Reading Bergson and Barad Diffractively." *Hypatia* 26, no. 1 (2011): 22–42.

Vaughan, Naomi E., and Timothy M. Lenton. "A Review of Climate Geoengineering Proposals." *Climatic Change* 109, nos. 3–4 (December 2011): 745–90.

Vernadsky, Vladimir I. "The Biosphere." In Samson and Pitt, *Biosphere and Noosphere Reader*, 35–38.

———. "Geochemistry." In Samson and Pitt, *Biosphere and Noosphere Reader*, 26–27.

Walker, Kenny, Jennifer Malkowski, and Damien Smith Pfister. "A Choreography of Living Texts: Selections from the ARST Oral History Project." *Rhetoric Review* 33, no. 3 (2014): 262–80.

Walsh, Brian, Philippe Ciais, Ivan A. Janssens, et al. "Pathways for Balancing CO_2 Emissions and Sinks." *Nature Communications* 8, no. 14856 (2017): 1–12.

Watts, Vanessa. "Indigenous Place-Thought and Agency amongst Humans and Non-Humans (First Woman and Sky Woman Go on a European World Tour!)." *Decolonization: Indigeneity, Education & Society* 2, no. 1 (2013): 20–34.

Weart, Spencer. "The Carbon Dioxide Greenhouse Effect." Discovery of Global Warming. Accessed February 19, 2019. www.aip.org.

Weier, John. "John Martin (1935–1993)." Earth Observatory (NASA). July 10, 2001. https://earthobservatory.nasa.gov.

Weinberg, Alvin M. "Can Technology Replace Social Engineering?" *Bulletin of the Atomic Scientists* 22, no. 10 (1966): 4–8.

Weiss, Joseph. *Shaping the Future on Haida Gwaii: Life beyond the Settler Colonialism*. Victoria: University of British Columbia Press, 2018.

Wells, Justine, Bridie McGreavy, Samantha Senda-Cook, and George F. McHendry Jr. "Introduction: Rhetoric's Ecologies." In McGreavy et al., *Tracing Rhetoric and Material Life*, 1–38.

White, Kyle Powys. "Indigeneity in Geoengineering Discourses: Some Considerations." *Ethics, Policy and Environment* 21, no. 3 (2018): 289–307.

Wibeck, Victoria, Anders Hansson, Jonas Anshelm, et al. "Making Sense of Climate Engineering: A Focus Group Study of Lay Publics in Four Countries." *Climatic Change* 145, nos. 1–2 (November 2017): 1–14.

Wiener, Norbert. *Cybernetics: Or Control and Communication in the Animal and the Machine*. Cambridge, MA: MIT Press, 1965.

Wild, John. "Plato's Theory of Texnh: A Phenomenological Interpretation." *Philosophy and Phenomenological Research* 1, no. 3 (March 1941): 255–93.

Wingenter, Oliver W., Karl B. Haase, Peter Strutton, et al. "Changing Concentrations of CO, CH_4, C_5H_8, CH_3Br, CH_3I, and Dimethyl Sulfide during the Southern Ocean Iron Enrichments Experiments." *PNAS* 101, no. 23 (2004): 8537–41.

Wilson, Sheena, Adam Carlson, and Imre Szeman. "On Petrocultures: Or, Why We Need to Understand Oil to Understand Everything Else." In *Petrocultures: Oil, Politics, Culture*, edited by Sheena Wilson, Adam Carlson, and Imre Szeman, 3–19. Montreal: McGill-Queen's University Press, 2017.

Yang, Andrew. "It's Worse Than You Think—Lower Emissions, Higher Ground." August 28, 2019. https://2020.yang2020.com.

Yusoff, Kathryn. "Anthropogenesis: Origins and Endings in the Anthropocene." *Theory, Culture and Society* 33, no. 2 (March 2016): 3–28.

———. *A Billion Black Anthropocenes or None*. Minneapolis: University of Minnesota Press, 2018.

———. "The Geoengine: Geoengineering and the Geopolitics of Planetary Modification." *Environment and Planning A: Economy and Space* 45, no. 12 (December 2013): 2799–808.

———. "Geosocial Strata." *Theory, Culture and Society* 34, no. 2–3 (March–May 2017): 105–27.

Zalasiewicz, Jan, Mark Williams, Will Steffen, and Paul Crutzen. "The New World of the Anthropocene." *Environmental Science and Technology* 44, no. 7 (2010): 2228–31.

Zerefos, C. S., V. T. Gerogiannis, D. Balis, and A. Kazantzidis. "Atmospheric Effects of Volcanic Eruptions as Seen by Famous Artists and Depicted in Their Paintings." *Atmospheric Chemistry and Physics* 7, no. 15 (2007): 4027–42.

Zielinski, Siegfried. *Deep Time of the Media: Toward an Archaeology of Hearing and Seeing by Technical Means*. Translated by Gloria Custance. Cambridge, MA: MIT Press, 2006.

INDEX

Page numbers in italics refer to figures and tables.